Electronic Warfare

Electronic Warfare

*From the Battle of Tsushima
to the Falklands
and Lebanon Conflicts*

Mario de Arcangelis

BLANDFORD PRESS
POOLE · DORSET

First published in the UK 1985 by Blandford Press Ltd,
Link House, West Street, Poole, Dorset, BH15 1LL

This edition copyright © 1985 Mario de Arcangelis

Originally published in Italian
by Mursia as La Guerra Elettronica:
Dalla battaglia di Tsushima ai giorni nostri.
Copyright © 1981
Ugo Mursia Editore,
Milan, Italy

Distributed in the United States by
Sterling Publishing Co., Inc.,
2 Park Avenue, New York, NY 10016

ISBN 0 7137 1501 4

British Library Cataloguing in Publication Data

De Archangelis, Mario
    Electronic warfare : from Tsushima to the
    Falklands and Lebanon conflicts.
    1. Military art and science—Automation—
    History
    I. Title   II. La guerra elettronica. *English*
    355.8   U104

Set in 11/12pt Monophoto Ehrhardt by
August Filmsetting, Haydock, St. Helens.
Printed in Great Britain by Biddles Ltd, Guildford.

Contents

Acknowledgements

Grateful acknowledgement is made first of all to Elettronica SpA., a true fountain of knowledge regarding electronic counter-measures, for their continous and generous assistance. I would particularly like to thank the company's Chairman, Ing. Filippo Fratalocchi, an expert on all matters concerning electronic warfare, for all that he has taught me. Warm thanks also to the company's Managing Director, Ing. Enzo Benigni, a courageous, modern entrepreneur, for his help and encouragement at all times.

Finally, I would like to extend my thanks and appreciation to my friend Miles Donnelly, with his broad international knowledge of the subject, for his precious collaboration and to Carol Preston for her excellent translation of this book for this present English edition.

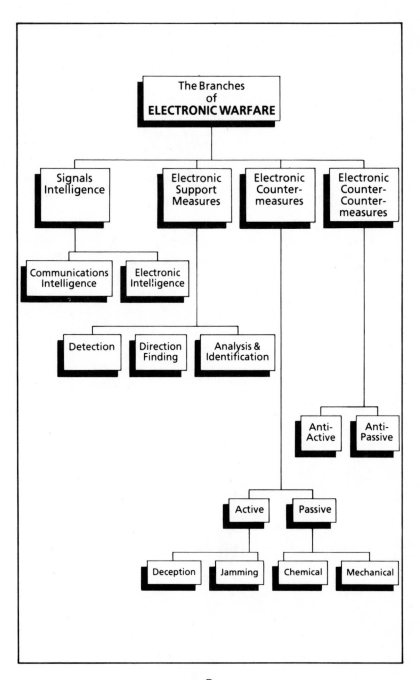

The Branches
of
ELECTRONIC WARFARE

Signals Intelligence

Electronic Support Measures

Electronic Counter-measures

Electronic Counter-Counter-measures

Communications Intelligence

Electronic Intelligence

Detection

Direction Finding

Analysis & Identification

Anti-Active

Anti-Passive

Active

Passive

Deception

Jamming

Chemical

Mechanical

Introduction

Nowadays, virtually everybody has some idea of what actually takes place in fighting between aircraft, tanks, fighter aircraft, and submarines. These vehicles are familiar to us, and everybody has seen them in action, either directly or via films and television. But the mention of electronic warfare usually conjures up rather vague notions of a special kind of conflict going on in the atmosphere and involving radio or radar emissions. What actually takes place in electronic warfare? What is this mysterious activity about which so much is talked and which goes on uninterrupted even in the quietest moments of peacetime?

And what are electronic countermeasures? Few people could give an acceptable answer to this question. Yet, together with electronic counter-countermeasures, they have played a key role in the latest conflicts in the Middle East and Southeast Asia. And electronic espionage? What is it? How is it carried out?

Electronic warfare has been one of the best kept secrets on both sides of the 'Iron Curtain' since the end of World War Two. Knowledge of matters concerning electronic warfare has always been the sole province of two extremely limited categories of persons: technical experts and the military forces. It has always been, and still is, in the interests of both these groups, though for different reasons, to keep the activity of electronic warfare hidden from indiscreet or suspicious eyes. For the crew of a military aircraft, a warship or a tank, an appropriate electronic countermeasure which has been kept secret can mean the difference between the success and failure of their mission, even the difference between life and death.

Consequently, there are strong reasons for keeping many aspects of electronic warfare secret. However, there are equally strong reasons for not only the military forces and those concerned with national defence being informed about the existence and general usefulness of electronic warfare, but the general public as well.

'If there is a World War Three, the winner will be the side that can best control and manage the electromagnetic spectrum'.

Admiral Thomas H. Moerer,
Former Chairman of the US Joint Chiefs of Staff.

1

The Origins of Electronic Warfare

The Russo-Japanese war, which broke out in February 1904 as a result of conflicting interests between St Petersburg and Tokyo, was the first war in which radio, or wireless telegraphy as it was called in those days, was used by both sides to communicate with their respective forces.

Wireless telegraphy had been invented by Guglielmo Marconi only a few years earlier but it had been immediately made use of, mainly by the naval forces, for long distance communications between ships, and between ships and land. The Japanese had installed wireless sets on all their ships; these were copies of Marconi's original invention but their performance was decidely inferior as they could operate on only one frequency and had a range of barely 60 miles. The Russians, too, had wireless sets on board their warships in the Far East and in numerous ground stations situated near their naval bases.

Right from the beginning of the war, the Russians used radio not only for normal communications but also, albeit in a somewhat improvised way, for purposes other than that for which it had been invented. These uses of radio could be considered to be the embryonic stage of electronic warfare. For example, Japan had started the war with a surprise attack on the Russian warships anchored in the ports of Chemulco and Port Arthur on the west coast of the Korean peninsula in the Yellow Sea, but, during the frequent subsequent Japanese attacks on Russian ships at Port Arthur, radio operators at the Russian base often noticed that, before an attack, they could hear in their headphones a great exchange of signals, increasing in intensity, between the Japanese ships; this was possible because the Japanese were using radio without taking any precautions to conceal transmissions. Since these signals were intercepted long before the enemy ships were sighted, the Russians were warned of the imminent attack and could therefore alert their ships and coastal batteries before the Japanese started their bombardment.

On one particular occasion, several Russian ships left the port of Vladivostok to make a surprise attack on the Japanese naval base of Gensan in the Sea of Japan, but the Japanese had discovered their departure and were waiting for them. However, as the Russian ships drew closer and closer to Gensan, they intercepted radio communications, increasing in intensity, which indicated the presence of numerous Japanese warships also bound for Gensan. The Russians, therefore, abandoned their plans, which would doubtlessly have ended in disaster since the entire enemy fleet was waiting for them at Gensan.

These were not the only occasions on which the Russians used radio for a purpose other than that of telecommunications in the first year of the war. On 8 March 1904, the Japanese attempted to carry out an attack on Russian ships anchored in the inner roads of Port Arthur, and thus not visible from the open sea. They sent two armoured cruisers, *Kasuga* and *Nisshin*, to bombard the roads by indirect fire, using a small destroyer favourably located nearer the coast to observe where the shells fell and to transmit correct firing instructions to the cruisers. However, a wireless operator at the Russian base heard the signals the Japanese ships were exchanging and, although he did not really understand what he was doing, he instinctively pressed the signalling key of his spark transmitter[1] in the hope that this might interfere in some way with the communications between the enemy ships. As a result of his instinctive action, no Russian ships were damaged by Japanese naval bombardment that day since the Japanese, their communications jammed, cut short their action and withdrew.

However, it was exploitation of the potential of radio by the Japanese, combining with a refusal by the Russians to make use of that potential, which brought the Russo–Japanese war to a conclusive end. Naval operations in 1904 were unfavourable to the Russians who, in the various battles with the Japanese fleet, lost most of their warships stationed in the Far East. For this reason, Russian leaders in St Petersburg decided to send the Baltic Fleet to the Far East to replace the lost ships and to seek revenge for the defeats that they had suffered. Admiral Zinoviy Petrovitch Rozhestvenskiy, who was to become the leading figure in one of the most dramatic events in the whole of naval history, was chosen to command the fleet.

Two years previously, in July 1902, Rozhestvenskiy, not yet an admiral, was in command of one of the thirty-one Russian warships, the cruiser *Ninin*, gathered in the roads of Reval in the Baltic Sea to greet Kaiser Wilhelm II of Germany who had come on his yacht to pay a visit to Tsar Nicholas III. After the prescribed gun salute in honour of the guest, the two Emperors and their entourage of ministers and admirals boarded *Ninin* to watch the Russian fleet perform a series of sea exercises.

These exercises, which consisted mainly of manoeuvres and firing at moving targets, went on for over three hours, throughout which Rozhestvenskiy, as though unaware of the presence of the two sovereigns on board his ship, conducted the manoeuvres with extreme calm and the utmost assurance. Rozhestvenskiy made such a favourable impression on the Kaiser that, upon disembarking, he congratulated the Tsar with the following words, 'I would be glad to have in my Navy officers as efficient as your Rozhestvenskiy.' The Tsar was also very impressed by the impeccable conduct of Captain Rozhestvenskiy who, of that day, was assured of a brilliant career.

On 14 October 1904, accompanied by the hopes and prayers of all Russia, fifty-nine ships of the Baltic fleet, under the command of Admiral Rozhestvenskiy, weighed anchor from Liepãja (Libaua) in the Gulf of Finland and set sail for the distant port of Vladivostok on the east coast of Siberia. They reached the Atlantic, circumnavigated the African continent and, after almost two hundred days of travelling, in which they covered 18,000 miles and encountered all sorts of difficulties, they finally entered the East China Sea towards the middle of May 1905.

At this point, Rozhestvenskiy had to decide which route they would take to enter the Sea of Japan and reach the port of Vladivostok. There are three means of access to the Sea of Japan: the Straits of Korea with the island of Tsushima in the middle, the Straits of Tsugaru between the two Japanese islands of Honshu and Hokkaido and, further north,

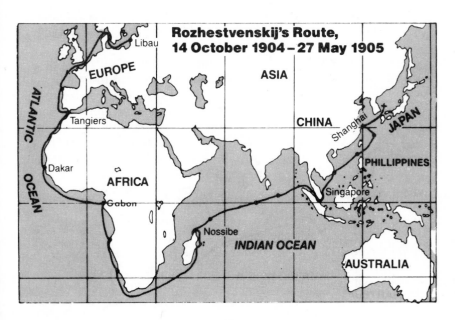

the Straits of La Pérouse between the island of Sakhalin and the northern end of the Japanese archipelago.

The choice of route was of the utmost importance since the destiny of the fleet might depend on it. The problem was how to reach Vladivostok and avoid encountering the Japanese fleet, given the precarious condition of efficiency after the long voyage. In fact, the question of which route to follow was highly debatable as each of the three routes had its advantages and disadvantages. For days, officers and sailors on board the ships talked of nothing else. Many were convinced that they should pass through one of the northern straits, Tsugaru or La Pérouse, since they were both a good distance away from the Japanese naval bases in Korea and were both closer to Vladivostok. This conviction also stemmed from the fact that, before leaving Russia, an extremely powerful radio apparatus had been installed on one of the ships, the auxiliary cruiser *Ural*. This apparatus had been specially built in Germany and had a range of almost 700 miles, truly exceptional for those days. The Russian sailors imagined that, with such an apparatus, they would be able to communicate with the remaining Russian ships at Vladivostok, getting them to come out of the port at the right moment, thus trapping the Japanese fleet in the cross-fire between the two Russian units.

The only person on board who never talked to anyone, not even his own staff, about the problem of selecting the route, was Admiral Rozhestvenskiy, possibly because he had already made his decision and had no intention of discussing the matter.

The Japanese fleet, under the command of Admiral Togo, was almost entirely concentrated in the bay of Mesampo at the southern end of the Straits of Korea, ready to come out into the open sea and intercept enemy ships on sight. However, the Japanese admiral had set up a system of surveillance consisting of a continuous patrol of carefully positioned ships. An old battleship was stationed south of the island of Tsushima as a relay point between patrol ships at sea and the Naval Command Squadron in port. The success of Togo's plan relied on the premise that he would have maximum advance sighting and swift early warning by radio. In short, the whole plan was based on the efficiency and speed of his radio communications network, without which the enemy could slip through.

On the other hand, the Russian admiral having weighed the pros and cons of the use of radio during his long voyage, had decided to dispense with this precious means of communication altogether. He reasoned that his main objective was to reach the port of Vladivostok without being discovered and attacked by the Japanese and, since Japanese interception of eventual radio communications would reveal the position of the Russian fleet, he ordered a complete radio blackout.

On 25 May 1905, the Russian fleet was steaming in two long lines at a speed of 9 knots, heading for the Straits of Korea. The sea was rough and visibility was poor. Early in the morning, the Russian fleet began to intercept weak signals on their radios. As they sailed further north, the signals increased in intensity and it was obvious that these communications were signals from the various Japanese patrol ships relaying to their squadron command.

Rozhestvenskiy appeared to ignore the existence completely of the enemy and, not even bothering to send out torpedo boats on reconnaissance, continued to steam ahead on his chosen route.

On the night of 27 May, with a thick mist and only a quarter moon, visibility was down to just over a mile. Nothing was sighted until about 02.45 when the cruiser *Shinano Maru*, patrolling at about 40 miles from the Goto islands, suddenly saw through the mist a ship steaming with her navigation lights on. The Japanese cruiser, unable to distinguish what type of ship she was, or her nationality, or whether she was alone or part of a formation, began to follow her without transmitting any radio messages regarding the sighting.

Later in the morning, at about 04.30, the *Shinano Maru* moved in closer and saw that the ship was a Russian hospital vessel. At this point, the latter noticed the presence of the Japanese cruiser and, mistaking her for a friend, flashed her signal lights. This error committed by the Russian ship led the commander of the *Shinano Maru* to infer that she must be part of a formation and so he moved in closer to investigate.

At about 04.45, the mist lifted and the *Shinano Maru* could make out a long line of Russian battleships and cruisers at a distance of just over half a mile from the hospital ship.

The *Shinano Maru* immediately started to radio the news to Admiral Togo's flagship but, given the distance between the two points and the prevailing atmospheric conditions, the primitive radio equipment on board was unable to communicate its precious message. Meanwhile, Rozhestvenskiy's ships had also spotted the ship which was now sailing on a course parallel to that of the Russian formation, disappearing every now and then in the morning mist. Although the Russians were not able to identify the ship, her actions made it clear that she was a patrolling enemy warship. Everyone expected Rozhestvenskiy to send out his fastest cruisers to destroy the imprudent enemy ship. It was an extremely critical moment as the destiny of the Russian fleet and the outcome of the whole war could depend on this decision. Rozhestvenskiy ordered the fleet to train all guns on *Shinano Maru*, but he gave no order to open fire.

Meanwhile, many Russian ships were intercepting the radio alarm of the *Shinano Maru*, calling the Japanese flagship. Aboard the *Ural*, equipped with the powerful, long range radio apparatus, the captain, indignant that no action was being taken against the Japanese patrol ship which by now seemed as if she wished to challenge the whole fleet, consulted his radio operator about the possibility of jamming the *Shinano Maru*'s transmissions. They both agreed that, if they transmitted a continuous signal on the same frequency as that used by the Japanese ship, this would disturb the latter's transmissions enough to prevent them from communicating their sighting of the Russian fleet. The captain duly signalled to the flagship requesting permission to use his radio for jamming purposes. After a few minutes, the Admiral replied laconically 'Don't prevent the Japanese transmitting.'

Rozhestvenskiy refused advice which, in those circumstances, might have proved invaluable. The motive behind his refusal is not clear; perhaps he wanted to prove to the fleet his self-confidence in the face of the enemy or perhaps he had failed to grasp the value of electronic jamming as a means of preventing the enemy from communicating.

Meanwhile, the *Shinano Maru*, without losing sight of the enemy, withdrew in order to ascertain the exact composition of the Russian fleet and to better observe its movements. Finally, radio contact was established and the 'enemy sighted' message was sent. She went on, uninterrupted, to transmit information regarding the course, position, speed and so on, of the enemy formation, which all clearly indicated that the Russians were heading for the Straits of Korea.

Just before dawn, a thick mist settled on the seas, providing the Russians with an excellent opportunity to slip away from the Japanese ship and head for the northern straits of Tsugaru or La Pérouse. Members of Rozhestvenskiy's staff begged him to re-examine the situation as it had become quite clear that radio contact had been established between the Japanese command and the *Shinano Maru*, which, by this time, had been joined by other patrol ships. But all efforts to persuade the Admiral were useless. At this point, exasperated by his stubborn refusal, senior officers of the fleet ordered their radio operators to disturb communications between the enemy ships in any way they could, but it was too late.

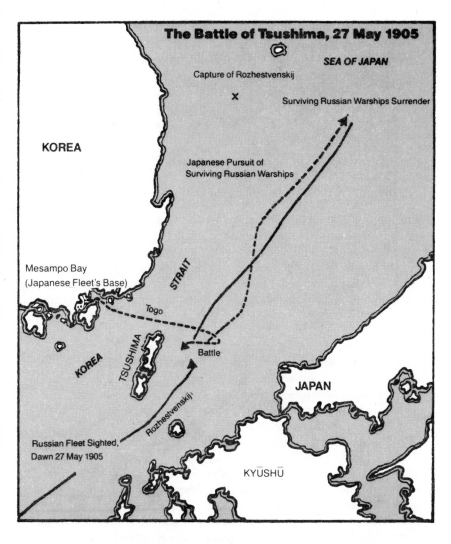

The Battle of Tsushima, 27 May 1905

SEA OF JAPAN

Capture of Rozhestvenskij

X

Surviving Russian Warships Surrender

KOREA

Japanese Pursuit of
Surviving Russian Warships

Mesampo Bay
(Japanese Fleet's Base)

STRAIT

Togo

KOREA

TSUSHIMA

Battle

JAPAN

Rozhestvenskij

Russian Fleet Sighted,
Dawn 27 May 1905

KYŪSHŪ

When the early dawn mist lifted, the scene was unchanged. Rozhestvenskiy's fleet, shadowed by enemy patrol ships, continued on course for the Straits of Korea, oblivious of everything, as though drawn towards an inescapable destiny.

The Japanese fleet was at sea. Admiral Togo, who had been anxiously awaiting news of the Russian fleet, had been extremely relieved to receive the *Shinano Maru*'s radio communication and immediately ordered his fleet to weight anchor and sail towards the enemy. At approximately 13.30, while the Russian formation, led by the battleship *Suvorov* with Rozhestvenskiy on board, was steaming into the Straits of Korea to the east of the island of Tsushima, the Japanese fleet suddenly appeared from over the horizon.

Rozhestvenskiy gave immediate orders to open fire. Two minutes later, as soon as his ships were in firing position, Admiral Togo replied with his guns. An avalanche of shells landed on the Russian flagship. The bridge, in which the whole Russian Staff was assembled, suffered repeated hits. Rozhestvenskiy himself was severely wounded and lost consciousness while his entire staff were either killed or wounded.

The outcome of the battle is well-known: Admiral Togo, with fast and brilliant manoeuvres, forced the Russian ships into a deadly cross-fire, inexorably destroying them one by one. Only three ships managed to reach Vladivostok while the rest of the fleet had to raise the white flag in surrender. On one of these ships was Admiral Rozhestvenskiy, wounded and unconscious, who was taken prisoner by the Japanese. That self-assurance and heedlessness of others which had so impressed the two Emperors during the Baltic Sea exercises proved fatal for Zinoviy Petrovitch Rozhestvenskiy when confronting the enemy in battle.

It is difficult to assess whether a different handling of operations would have yielded better results for the extremely difficult mission that was assigned to Admiral Rozhestvenskiy. Even if the *Shinano Maru*'s radio messages to Admiral Togo had been jammed, thus depriving him of knowledge of the Russian fleet, the superiority of the Japanese fleet might conceivably have secured the victory in any case. However, one cannot fail to observe that, if there had been a way for Rozhestvenskiy to prevent the deaths of many thousands of Russian sailors, it was by the prompt use of rudimentary electronic counter-measures. The *Ural*'s powerful radio transmitter could have been used, as her captain had urged, for jamming which would have prevented, or at least slowed down, enemy communications to Admiral Togo regarding their sighting of the Russian fleet.

2

The Dawn of Electronic Warfare

The Austrians were the first to realise that intercepting radio transmissions was an excellent means of acquiring political and military intelligence previously sought only through costly and dangerous espionage undertakings. In fact, when a political crisis arose with Italy in 1908 as a result of the Austro-Hungarian Empire's annexation of Bosnia and Hercegovina, the Austrians intercepted and deciphered Italian radio traffic and used the electronically-acquired intelligence to shape their foreign policy.

In 1911, during the Italo-Turkish war, the Austrians gave another demonstration of the capability of their intelligence service. Still extremely interested in Italian political and military affairs, the Austrians intercepted every radio message transmitted between Rome and Tripoli, where the Italians had landed, which provided them with information about the movements of the troops and the daily combat situation.

This was doubtlessly the first time in history that technical means (radio), instead of traditional means, such as spies or cavalry scouts, had been used to follow, step by step, the course of a campaign conducted hundreds of miles away.

Another nation which, like Austria, had always cultivated the art of espionage in warfare, was France. In the years preceding World War One, the French intelligence service managed to intercept and record all the transmissions made by foreign embassies in Paris to their respective governments and all diplomatic messages coming from abroad.

An outstanding example of French electronic espionage was the interception of a long message transmitted to the German ambassador in Paris from the German Foreign Minister containing a declaration of war to be delivered to the French government. The French, who had already cracked the code in which the message was sent, not only intercepted the despatch but so garbled its contents that the German amabassador could at first make nothing of it, while the French gained valuable time in which to prepare for mobilization.

During World War One the interception of diplomatic radio traffic reached incredible proportions. The British intelligence services broke the top secret German codes and, for three years, were able to intercept and decipher all the messages that the German Foreign Ministry sent to its embassies abroad. The British managed to keep this secret and only mentioned it to their American allies when the Germans, who were totally unaware of this leak in their intelligence service, tried to entice Mexico into the war with the promise of the annexation of Texas, Arizona and New Mexico.

However, from the electronic point of view, World War One is to be remembered mainly for some important events which can be considered the true beginning of electronic warfare.

In 1914, just after Great Britain had declared war on Germany, a remarkable incident took place in the Mediterranean. The German cruisers *Goeben* and *Breslau* were being shadowed by the British cruiser *Gloucester*, which was under instructions to radio all movements of the German ships to the Admiralty in London. The Admiralty would then send orders to the Mediterranean Fleet to intercept and destroy the two German cruisers: unfortunately, however, the British had no idea which route the cruisers would take since they could head for Italy, at that time neutral, or a friendly Turkish port. Radio communications between the *Gloucester* and the Admiralty were intercepted by the German cruisers who, at an opportune moment, decided to shake off their pursuers by disturbing their communications. They did this by transmitting a chaotic noise on the same frequency as that used by the British. The British changed the frequencies of their transmissions several times but to no avail. The German ships suddenly changed course and headed at full speed for the friendly waters of the Turkish Dardanelles.

This communication jamming could be considered the first real action of electronic warfare since, for the first time in history, electromagnetic waves had been used, not for communication, but for jamming enemy communications.

Electronic warfare was also engaged, although in a less obvious way, on the home fronts. A few years before the outbreak of World War One, Austria and France had set up special units to intercept army radio traffic. Germany did not set up such a system until several months after the war had begun, strangely enough, so Austria furnished the German intelligence service with precious information gleaned from intercepted enemy communications. To be fair, many other nations as well as Germany, had been slow to grasp the importance of intercepting enemy radio transmissions.

The Russians—despite their experiences in 1904—were particularly uninformed and naive regarding the use of radio. At the beginning of the war, they did not even seem to realise that radio transmissions could be received by anybody who happened to be listening on the same frequency. German interception of messages transmitted in plain Russian contributed greatly to General Hindenburg's victory over the Russians in the Battle of Tannenberg. Later, the Russians realised that it was necessary to send their messages in code but the skilful Austrians were able to break the code promptly and decipher their messages. The Germans, therefore, received daily information regarding Russian movements on the Eastern front right up to the Bolshevik revolution in 1917 and the Armistice.

The French were also well-organised in these activities and right from the beginning of World War One were able to intercept and decipher radio communications transmitted by the Germans who, like the Russians on the Eastern front, inexplicably committed various serious errors in their employment of radio.

By now, all the various military leaders and their respective staff were beginning to understand and appreciate the operational advantages to be gained from intercepting enemy communications and were requesting greater support in this new field. And so electronic espionage was born, an activity which was to play an increasingly important role in modern warfare.

Although barely fifteen years had passed since Guglielmo Marconi had invented it, radio had already evolved to the stage where it could be effectively employed on ships, aircraft and in both fixed and mobile ground stations. This had been done right from the beginning of World War One but it soon became apparent that electronic espionage required equipment that was more sensitive than the receivers then in common use.

The aim was to intercept, record and analyse all the messages transmitted by the enemy, whether in plain text or in code, including those which were barely perceptible. Code-breakers were employed to decipher coded messages. For them to be able to do this, it was necessary to intercept a large number of coded enemy messages. Statistical techniques, such as counting the number of times characteristic phrases like 'in answer to' or 'nothing new' were used, provided data which was extremely useful in breaking the enemy's code. However, it was not always necessary to decipher the whole of an enemy coded message in order to glean the essential information. Vital information regarding the enemy's position and intentions could nearly always be obtained from a first analysis of radio traffic. To improve reception of enemy communications, receiving sets were equipped with amplifiers using a device invented only a few years previously, the electronic tube or amplifying valve.

In order to intercept enemy communications, the first thing to do is, obviously, to find the frequency on which the enemy is transmitting. Since, in wartime, this is often changed in an attempt to maintain secrecy, great patience was required from the highly-skilled operators who spent hours tuning their sets until they found the frequency being used by the enemy. Once the frequency had been located, all radio traffic was received and recorded until the enemy changed frequency.

In the first few years of World War One the frequencies generally used for radio communications were between 150 and 750 Kc/s. It was known that frequency determines many aspects of radio transmission, mainly relating to range, but also that, the higher the frequency, the smaller could be the components of the radio set. In other words, the performance and dimensions of such sets depended also on the frequency used. Thus, in many cases, high frequencies were used so that smaller sets could be built to install in aeroplanes, for example. Towards the end of World War One, the frequencies used were between 750 Kc/s and 1Mc/s (one million cycles per second), tending to increase as each side tried to make it more difficult for the enemy to intercept their transmissions.

During World War One, both sides also experimented with electronic deception in its simplest forms such as false transmissions, dummy traffic and other similar ruses for misleading the enemy.

Wire communications were also susceptible to interception by the enemy. On the home fronts, the telephone was in general use and the most elaborate stratagems were devised in order to eavesdrop on enemy communications. During trench warfare, the forces mostly adopted telephonic systems using single-wire ground-return lines. Since the single wire was on home ground and military leaders were convinced that the enemy could only listen in on their conversations by getting onto the line itself, they were not the least bit worried about being intercepted and so took no precautions. This conviction turned out to be completely unfounded and the first to learn the lesson was the British Expeditionary Force in France who, as early as 1915, began to realise that the Germans were managing to foresee and prevent their operations with annoying regularity. It was almost as though the Germans had received a copy of the orders relating to planned British offensives!

In fact, the Germans had built an apparatus which, by means of a network of copper wires or metallic rods buried as near as possible to enemy lines, was able to pick up even the faintest current created by the British telephonic system's ground return. The stray ground currents and leaks were picked up and amplified, using a recently invented, extremely sensitive amplifying tube. Thus, the Germans were able to exploit the enemy's indiscriminate use of the telephone by intercepting their communications through the ground.

As soon as this ingenious system was discovered, the British immediately came up with an antidote—an apparatus capable of blocking the diffusion of sound through the ground within a certain radius from the source of emission. This device not only put an end to the enemy's interception of telephone communications, but also led to the development of a new system for intercepting telephone communications through the ground. This new system, in use by the following year, utilized a greater number of electronic tubes and other sophisticated technical devices and was capable of achieving telephonic intercepts up to a distance of 4–5000 metres.

During the last two years of the war, these systems of telephonic interception had become so effective that the various military leaders on the Western Front, realising the disadvantages of the telephone, restricted its use considerably.

Since the beginning of the war military engineers and technicians had been devoting their efforts to the construction of more sophisticated equipment, not only to improve communications among their own units, but also to discover and locate the enemy's radio stations. This was made possible by the development of a radio-goniometric system designed by the Italian scientist, Professor Artom, who had discovered the 'orienting' action of the loop antenna; that is, the ability of such an antenna to establish the direction from which an electromagnetic emission originates.

Artom's antenna was used in the Bellini-Tosi direction-finder (DF) which consisted of two crossed loops and was ideal for discovering the direction of radio stations transmitting on medium-wave and long-wave frequencies. Guglielmo Marconi, who had moved to England a few years previously, perfected the method invented by his compatriot Artom using a new, extremely sensitive amplifier tube capable of picking up even the faintest signals which the normal crystal radio sets had been unable to detect. As early as 1914, this new equipment made it possible to intercept enemy electromagnetic emissions and determine the direction from which they were coming, thus establishing the position of the transmitting station.

The direction-finder thus became a precious instrument in

electronic espionage, and for acquiring information about the enemy. The use of radio by the armed forces was not very widespread in those days so the localisation of an enemy transmitting station nearly always indicated the presence of a large military unit in the area; moreover, the territorial distribution of the radio stations gave a very clear idea of organisation on the enemy front and changes in location of a radio station provided a fairly exact picture of troop movements. The French and British were particularly well-organised in this field and since 1915 had employed effective radio-goniometric systems of interception which enabled them to establish the location of large enemy units, movement of troops and plans of attack. All this contributed greatly to the success of the Allies in wearing down the enemy and forcing them into a static position with a high attrition rate.

The direction-finder achieved its greatest success in maritime operations during World War One. The British, in particular, scored outstanding successes in determining the movements of German submarines, which were obliged to surface to transmit information to their commands. Many submarines sunk in those days could be credited to the British direction-finding system which supplied the anti-submarine warships with information regarding the movements of enemy submarines. Actually, it was not very difficult for the British to acquire such information as the German submarines used radio without taking any precautions. Equipped with powerful transmitters operating on a frequency of 750 Kc/s, the German submarines surfaced at fixed times to transmit long messages to their commands. These messages were rather stereotyped which greatly facilitated the work, not only of the codebreakers, but also of the British DF operators who had to determine the direction of the transmissions and the exact position of the submarines themselves.

Technical progress in the field of radio and its accessories made it possible to build smaller, lighter direction-finders which could be transported by undercover agents. This development was exploited by the Germans in their airship raids on England.

When the Germans started bombing London at night, they realised that they would have to solve the problem of how to reach the target in the dark. At first the German airships used astral navigation systems, but these proved unsatisfactory due to the intrinsic unsuitability of the airship, and to atmospheric conditions such as fog and clouds. So the Germans abandoned these systems and adopted instead a long-range radio-guidance system using a network of transmitters installed in Germany. However, this system also proved ineffective as the on-board receivers were not capable of sufficient accuracy either because of the great distance or as a result of errors caused by multi-path night fading.

Finally, the Germans sent undercover agents to England who installed themselves, with portable radio beacons, in a house just outside London. From there, they were able to guide the airships to their target with sufficient precision in spite of the dark or fog. But the presence in the air of strange electro-magnetic signals just before a bombardment soon roused the suspicions of the British Secret Service who, using direction-finders mounted on vehicles, began a systematic search for these emissions.

The German airships committed serious errors in their use of radio as, like the submarines, they always transmitted on the same frequency and always used the same code-names when communicating with ground stations. Moreover, they flew at a fairly low speed. All in all, it was fairly easy for the British to know when a raid on London was imminent. It was also quite easy for the British to locate the building used by the German agents, who were arrested. Instead of dismantling the clandestine station, however, the British used it the following night to guide the airships to an uninhabited area on the North Sea coast where British fighter aircraft were waiting for them; the result was total destruction of the enemy airships.

After this incident, airships were no longer used as bombers as it had become more than apparent to the Germans that they were exceedingly vulnerable to enemy fighters. Bomber aircraft, such as the Gotha, were sent to bomb London while the airships were relegated to perform subsidiary tasks.

The most interesting and successful operation ever carried out by the British direction-finder network was that which took place just before the great naval battle of Jutland. In 1916, British public opinion registered grave discontent regarding the passive conduct of the Grand Fleet, which had failed to impede the German Fleet's incursions in various coastal areas of Great Britain. The bitter memory of the battle of Dogger Bank, in which Admiral Hipper of the German navy had successfully eluded the actions of the British fleet commanded by Admiral Beatty, pained the very souls of those who felt they ruled the waves and they demanded revenge! However, the geographical situation, distances between bases and other relevant factors, all favoured the German fleet, which always managed to 'hit and run' before the British arrived; it was a problem of timing which was difficult to solve.

For the end of May of the same year, the Germans planned a major naval incursion against the British coastline in which submarines and airships would participate. To prevent the British direction-finder network from detecting the departure of their fleet from port, the Germans planned to deceive the British Admiralty by means of an electronic stratagem.

A few days before weighing anchor, the Germans exchanged the radio telegraphic code-name of their flagship the *Friedrich der Grosse* with that of a radio station at Wilhelmshaven, where the German fleet was based. In this way, the British, who regularly intercepted the flagship's transmissions, would think that the German fleet was still at Wilhelmshaven.

However, towards the end of May, British radio operators noticed a sudden increase in the number of transmissions made by an unknown ship in the port of Wilhelmshaven which requested canal mine-sweepers, fuel supplies, and so forth. These messages were a clear indication that the German fleet was preparing for an important sea operation, so all radio stations along the British coast were put on the alert to keep an eye on what was happening at Wilhelmshaven.

On 30 May the faith that the British navy had placed in its interception and direction-finding service paid off when changes in direction of the unknown ship's transmissions were noticed. These changes convinced the Admiralty that the German ship, and probably the whole fleet, had left its base and was once again planning to bombard a target in Great Britain. The Admiralty gave immediate orders to Lord Jellicoe, Commander-in-Chief of the Grand Fleet, to weigh anchor and set sail with all haste for the Gulf of Heligoland.

While the two fleets were sailing at top speed towards each other, the Germans sent their Zeppelin airships to explore the sea area to the west of the Danish peninsula. This reconnaissance proved fruitless for them but not, however, for the British ships whose direction-finding stations on the French coast picked up the airships' signals, thus confirming that the German fleet had, in fact, put to sea.

The outcome of all this was the battle of Jutland, one of the most important battles in naval history. Much has been written about this battle but possibly no-one has pointed out that it would probably have never taken place without the British interception and direction-finder service!

3

The Battle of the River Plate and the Advent of Radar

In 1939, just before the outbreak of World War Two, the German pocket battleships *Deutschland* and *Admiral Graf Spee* were ready and waiting in the Atlantic. They were fast ships with superior endurance, excellent armour and an impressive array of weapons including six 280 mm guns. Whilst packing a great punch, they had a displacement of only about 10,000 tons and, for this reason, were called pocket battleships. As they could hold their own against almost any naval vessel except a true battleship (a warship of approximately 35,000 tons), they had been selected to operate as raiders with the same hit and run tactics which had made other German ships so famous during World War One.

The *Deutschland*, operating in the North Atlantic, encountered no enemy merchant ships and, when her fuel started to run low, headed back to Germany via the Norwegian Sea. The *Graf Spee*, operating in the South Atlantic, sank nine British merchant ships; the last one, sunk on 3 December 1939, managed to communicate by radio that she had been attacked by the German ship halfway between the Cape of Good Hope and Sierra Leone.

The task of protecting Allied merchant traffic in that part of the Atlantic had been entrusted to a British naval force composed of the three cruisers *Ajax*, *Achilles* and *Exeter* under the command of Commodore Harwood. On receiving the radio communication from the sinking ship, Harwood presumed that the *Graf Spee*, having now been discovered by the enemy, would move out of her operating zone and would probably head for the busy estuary of the River Plate.

Calculating that he could get there himself in about ten days and, therefore, arrive at roughly the same time as the *Graf Spee*, Harwood immediately ordered his ships to set sail for the River Plate, maintaining absolute radio silence. At daybreak on 13 December, the *Graf Spee* was, in fact, sighted, interception of radio transmissions made by the German ship perhaps helping the British to find her.

Captain Langsdorff, in command of the *Graf Spee*, was confident in the superior armament of his ship and gave immediate orders to open fire, causing heavy damage to the British ships. *Graf Spee*, however, was also hit and had to take refuge in the neutral port of Montevideo for repairs.

The British ships followed her but waited outside the estuary knowing that she would have to leave after seventy-two hours, this being the maximum time a warship was permitted to remain in a neutral port according to The Hague International Convention. Captain Langsdorff asked for more time but his protests were in vain and, convinced that his ship was in no condition to confront the British, now reinforced? sailed out of territorial waters and ordered the ship to be scuttled. After making sure that all the crew had safely abandoned ship, Langsdorff, holding himself entirely responsible for the loss of the ship, committed suicide.

However, during the *Graf Spee*'s brief stay in the port of Montevideo, the British naval attaché there had taken several photographs of the ship which were promptly forwarded to the Admiralty in London. During these dramatic moments, the crew of the *Graf Spee* had neglected to cover the radar antenna which showed up clearly on the photographs taken by the British. When British intelligence technicians examined these photographs, they were extremely surprised to discover that the ship had radar equipment far superior to any British radar then in existence.

Electronic experts were sent to Montevideo to inspect the remains of the *Graf Spee*, hoping to find out, by examining the antenna, more about the type of radar the Germans had installed on board this ship. It was, in fact, the famous 'Seetakt' fire control radar which operated on a frequency of 375 Mc/s with a wavelength of only 80 cms. One of the first 'centimeter' radars, it was a very sophisticated piece of equipment characterised by remarkable precision over a range of more than 9 miles. What worried the British was that German technology seemed far more advanced than theirs, British industry not having yet produced a radar apparatus of this calibre.

Fortunately, however, only three sets of 'Seetakt' had yet been built by the Germans although the British did not know this! The 'Seetakt' had been able to detect the three British cruisers and, during the battle, had provided precise information regarding the distance of the targets until the radar had been hit by the British and put out of action. The *Graf Spee* incident, while demonstrating that Great Britain had little to fear from enemy 'raider' ships, also showed up the shortcomings of the radar equipment installed on Her Britannic Majesty's ships.

The British immediately began research on a naval radar to compete with the 'Seetakt' and began studying the possibility of neutralising it by appropriate electronic counter-measures. This was the first time in history that radar, the great secret weapon of World War Two, had been used in battle.

It is often thought that radar was a British invention, perhaps because the British were the first to use it systematically in the field of air defence. But, in fact, research was also being carried out simultaneously in Germany, Italy, France and the United States.

The general principles of radar had been formulated some time before and were known to all. In 1888, the German physicist Heinrich Herz had proved that electromagnetic waves, thereafter called 'Herzian' waves, behave like light-rays in that they can be channelled into a single beam and bounced off a metal surface, giving rise to a return echo which can be picked up.

A few years later, in 1904, an engineer from Dusseldorf called Christian Hulsmeyer requested a patent for a 'radiophonic measuring apparatus', which consisted of a transmitter and a receiver mounted side by side. The two devices were built in such a way that waves emitted by the transmitter, would activate the receiver if reflected by a metallic object. This apparatus, which the German engineer called a telemobilscope, was able to pick up sounds, like the chiming of a bell by receiving electromagnetic waves bounced off metallic objects at a distance of a few hundred yards. However, despite the success of an experiment carried out at Rotterdam, the big shipping companies showed no interest in Hulsmeyer's apparatus. It was perhaps too early for people to appreciate the potential value of such an apparatus. In fact, at that time, little was known about radio waves; there were no means of amplifying a signal, protecting it from external interference, or controlling the electromagnetic emissions produced, and so on.

Little technical progress had been made by 1922 when Guglielmo Marconi, during a conference held at the Institute of Radio Engineers in the USA, expounded practical ways of using radio waves for maritime navigation. He envisaged an apparatus capable of radiating a beam of electromagnetic waves in a fixed direction which, on meeting a metallic object such as a ship, would be bounced back.

In 1933, in the presence of Italian military authorities, Marconi demonstrated 'interference' in the reception of signals caused by motor-cars passing in the vicinity of a radio beam from a radio station, using a wavelength of 90 cms, linking Rome and Castelgandolfo.

Marconi's initiative resulted in a formal proposal, approved by the Italian Ministry of War in 1935, for the construction of a Radio-Detector Telemeter (RDT). Of the three armed forces, the Italian navy was the most interested in, and best equipped to deal with electronic research and development. A research project was therefore set up under the direction of Professor Tiberio at the Institute of Mariteleradar, in association with the Naval Academy of Livorno.

However, both finances and labour were in extremely short supply so Professor Tiberio, who had meanwhile been appointed a naval officer, had to develop the prototype almost single-handed. It was only in 1941, after the battle of Cape Matapan in which the Italian navy lost three cruisers, two destroyers and 2,300 men, that the authorities learnt that the British had electronic equipment for night-sighting on board their ships. The Italian Admiralty had received the impression, during the battle, that the British were using such equipment to direct their manoeuvres and firing; this was, in fact, confirmed by interception of a coded message from Admiral Cunningham, the commander of the British naval squadron. Immediately, the Italian authorities released sufficient funds for the completion of the 'Gufo' radar sets which were then still at the experimental stage at Livorno.

However, the most important contribution to the development of radar was made by two US physicists, Gregory Breit and Merle Tuve, in 1924. They set up experiments using radio pulses to determine the height of the layer of ionized gas which surrounds the Earth. By measuring the time these pulses took to reach the gas layer and return to Earth, they discovered that the ionized gas layer was at a height of about 70 miles and that it reflected radio waves.

In Germany, in the early 1930s, Doctor Rudolph Kunhold, head of the research division of the German navy, was trying to develop an apparatus capable of detecting objects underwater by bouncing sound-waves off them; this apparatus now goes by the name of sonar. In these experiments Doctor Kunhold realised that what was accomplished underwater could also be accomplished above it, using radio waves. He conducted several experiments in this new field and incorporated into his apparatus a new electronic tube capable of generating power of 70 watts on a frequency of 600 Mc/s—truly exceptional in those days. With this new electronic tube, produced by the Philips company of Holland, Kunhold completed the construction of his radar apparatus in 1934 in the research laboratories of the German navy at Pelzerhaken. The presentation of the new apparatus to high-ranking naval officers was a great success as, besides being able to detect a ship at a distance of 7 miles, the radar also located a small aircraft which happened to be passing by.

In the United States, radar research was being carried out both by the Signal Corps and the Naval Research Laboratory, which were working independently. In 1936, the Naval Research Laboratory developed a prototype operating on 200 MHz and the first series of these systems, bearing the trademark CXAM, were installed on board major naval units in 1941. In 1939–40, the Signal Corps developed a long-range system designated SCR-270. One such system was in operation at Pearl Harbor on the morning of 7 December 1941 but, although the radar operator received signals of approaching aircraft, nobody alerted the ships in port.

In Britain, studies in the field of short wave signals were initially undertaken for purely scientific purposes, such as determining the height of certain conducting layers of the ionosphere discovered by the British physicist E.V. Appleton (Appleton layers) in 1926. However, war clouds were gathering on the horizon, and the realization that Britain was particularly exposed to air raids, led to a drastic increase in scientific effort in an attempt to make up for lost time.

The first result of this effort was when the physicist Robert A. Watson-Watt, a descendant of the famous James Watt who gave his name to the unit of electrical energy, succeeded in visualising radio signals by means of Braun's cathode ray tube, and in determining electro-optically the emission propagation time. A few years later, in 1935, Watson-Watt developed the first practical equipment for detecting the presence of aeroplanes.

Radar is not considered to be an instrument of electronic warfare; it is, rather, the main target of electronic warfare, the enemy to be confronted. Radar is an electronic eye which can see in the dark and fog and which can penetrate smoke-screens. It can detect the approach of the enemy at much greater distances than the naked human eye can; it can direct gunfire in conditions of poor visibility and can even provide information regarding the topographical features of the zone.

A radar set consists of a transmitter, a receiver, an antenna and a screen, or radar scope. The transmitter sends out pulses of electromagnetic energy via a highly directional antenna pointing in a fixed direction. If a pulse meets a target, for example an aeroplane, during its journey, it is bounced back, or reflected, towards the receiver. The time-lapse between the transmission on the pulse and receipt of the return echo is measured by a special device incorporated into the radar set and, since it is known that electromagnetic waves travel at a speed of 300,000 kms per second, it is an easy matter to calculate the distance of the target. The operator can thus read directly on his radar scope both the distance and the bearings of the target.

4

The Sinking of the Bismarck

Contrary to the old Latin saying, fortune did not favour the brave German sailors during the memorable, dramatic adventure of the battleship *Bismarck* which took place in the Atlantic in May 1941.

The powerful German battleship, escorted by the 10,000-ton cruiser *Prinz Eugen*, left Bergen, Norway, on the evening of 22 May 1941, bound for the Atlantic where she was to join the battle cruisers *Scharnhorst* and *Gneisenau*, stationed at Brest, and with them operate as a raider group against British merchant ships.

The *Bismarck* group under the command of Admiral Lütjens left the Norwegian fjord and set sail for the Denmark Strait between Iceland and Greenland. The next day, however, a British reconnaissance aircraft reported the departure of the ships from the Norwegian fjord and the Home Fleet Command gave immediate orders to stop the German ships from entering the Atlantic.

The British cruisers *Norfolk* and *Suffolk* were patrolling the western exit of the Denmark Strait, operating as radar pickets.[1] The *Norfolk* was equipped with a 286 P-type radar, operating on a 1.5 metre wavelength, but with a fixed antenna which severely limited the search angle of the radar. The *Suffolk* was equipped with two radars; a 279-type operating on the same frequency as that of the *Norfolk* but with a rotating antenna particularly suitable for air search but also effective in the surface role, and a newer 284 MKV-type operating on a 50 cm wavelength with a 15-mile range, also fitted with a rotating antenna which, however, had a blind stern sector.

The two German ships, on the other hand, aside from being equipped with DeTe, 'Seetakt' radar, were also fitted with two receivers, a radio direction-finder (DF) receiver and a new piece of equipment which had the function of intercepting electromagnetic impulses emitted by an enemy radar. This new apparatus, made by Metox, picked up radar signals transmitted on frequencies between 110 and 500 Mc/s and was thus able to detect the presence of any ships, submarines, aircraft or other platforms which had a radar apparatus transmitting on these frequencies.

The great advantage of radar warning receivers (RWR) such as the Metox is that they are able to detect an enemy radar before that radar detects them; this is due to the fact that the RWR receives the direct

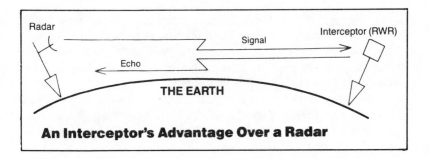

An Interceptor's Advantage Over a Radar

signal emitted by the radar while the latter picks up the signal only after it has been reflected by the target. This advantage means that, under normal conditions, the RWR's range is about one and a half times greater than that of radar and, in some cases, can be almost double.

The usefulness of such advanced electronic equipment was so highly regarded by the Germans that they had sent sets of Metox by submarine to both the cruiser *Hipper* and the pocket battleship *Admiral Scheer*, which were operating in the Atlantic. These sets, along with the requisite technicians and operators, were embarked on the ships while at sea and, once installed, the equipment enabled the ships to evade repeatedly the British warships which were hunting them and to return to their respective ports unharmed after sinking many enemy merchant ships.

On the evening of 23 May 1941, the *Bismarck* group entered the Denmark Strait. Admiral Lütjens, convinced that the British did not have a radar comparable to the 'Seetakt', was confident that they could get out of the Denmark Strait unseen, especially as conditions of visibility were poor.

The *Bismarck* group was obliged to follow a narrow course due to the presence of icebergs in the Channel which was therefore easier to cover with radar but, in spite of this advantage to the British, it was the Metox which first detected the presence of the enemy. With the aid of this electronic interceptor, the Germans were informed not only that British ships were waiting for them and that they had radar on board but also that it was a fairly advanced apparatus, judging by the frequency employed.

A few minutes later, the *Suffolk* managed to locate the enemy and both British ships steamed towards the enemy to attack. But as they emerged from the mist, they encountered, at a distance of only 5–6 miles, the German battleship lying in ambush. *Bismarck* loosed off a salvo of five rounds from her eight 381 mm guns at the leading ship *Norfolk* which wisely about-turned and moved back into the mist. She managed, however, to maintain radar contact, and transmitted a message that the enemy had been sighted.

As soon as the message was picked up, the battle cruiser *Hood* and the battleship *Prince of Wales*, which were closest to the position indicated, immediately headed at full speed towards the target while the entire Home Fleet, battleships, cruisers, destroyers and, above all, aircraft carriers, began to converge from all directions.

At dawn on 24 May the *Hood* established radar contact with the *Bismarck* group and, with the *Prince of Wales*, engaged the German ships in battle. At 05.52, the *Hood* opened fire at a distance of about 23,000 yards, immediately followed by her sister ship. At 05.55, the two German ships opened fire. The distance between the two contenders rapidly diminished and the *Bismarck* group's firing became more and more accurate. The third salvo hit the *Hood* squarely and the mighty battle cruiser, pride of the Royal Navy, shot into the air with a tremendous explosion: her back shattered, her stern and bow tilted up into the air and then disappeared beneath the waves, along with all but three of her crew. The *Bismarck* group then trained its guns on *Prince of Wales*, inflicting serious damage on the British ship, forcing her to retreat once again into the mist. *Bismarck* had herself been hit by the *Prince of Wales*, however, and was now leaking fuel.

The German admiral ordered the ships to proceed southwards at full speed and, since the Metox had received no enemy signal for the past six hours, decided that he could safely transmit by radio a description of the battle, informing the German supreme command of their victory. The transmission, which lasted over half an hour, gave the British a superb and unexpected opportunity to fix the position of the fleeing German ships via their direction-finding stations in Ireland and Gibraltar.

This was communicated to the *Norfolk* and the *Suffolk* who were hunting for the *Bismarck*, thus enabling them to make radar contact towards the evening. At about 18.00 the German ship, realising, thanks to the Metox, that she had once more been located by enemy radar, suddenly changed course and opened fire on the British ships who rapidly withdrew. This sudden action had the two results that Admiral Lütjens had foreseen: it shook off the pursuers and permitted the *Prinz Eugen* to break away from the *Bismarck* and sail alone to the French port of Brest.

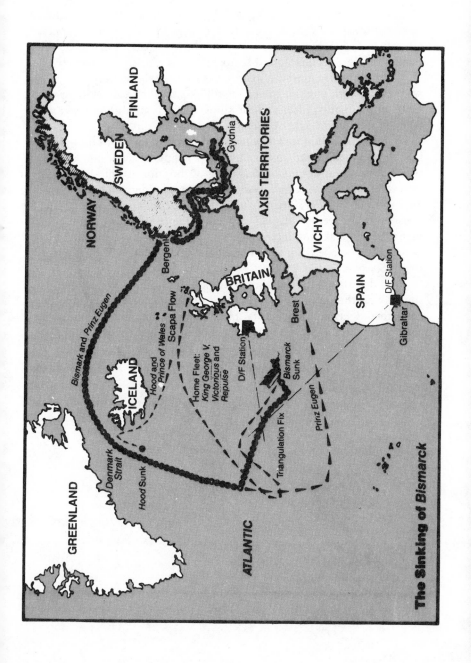

The Sinking of Bismarck

The *Bismarck* was next spotted by a Royal Air Force Catalina reconnaissance aircraft, and was then attacked by Swordfish torpedo aircraft from HMS *Victorious*, one of whose torpedoes damaged her rudder leaving her helpless. Meanwhile, other British ships arrived and kept her under constant fire until, at 10.40 on the 27 May 1941, she went down with most of the crew still aboard. The Commander-in-Chief of the British Fleet and the ships' crews praised the *Bismarck's* courage during her fight against such a superior force.

In the light of what we know today, however, it must be pointed out that it was the failure to observe a basic principle of electronic warfare which caused the sighting and sinking of the ship. Admiral Lütjen's decision to transmit a long report to the German High Command was an extremely serious error which had dire consequences. Strict observance of radio silence, especially since the formation had managed to shake of its pursuers, might well have enabled the *Bismarck* to escape from the Home Fleet and repair to Brest like the *Prinz Eugen.*

The defeat of the mighty German battleship, while greatly simplifying the problem of defending merchant traffic, nevertheless confirmed German superiority in the field of radar and the first electronic warfare systems (such as the Metox). The British, after this experience, spared no effort in order to bridge the technological gap in the field of applied electronics for military use.

5

The Birth of Electronic Countermeasures

By the summer of 1940, Germany had conquered nearly all of continental Europe and could now dedicate all her efforts to Great Britain, her old enemy! According to Hitler, the only way of dealing with the British was to invade their island. To this end plans were drawn up, code-named 'Sea Lion', in the form of a landing to take place around mid-September of that year.

The first step was to put the British Royal Air Force out of action. Then the *Luftwaffe* would keep the British Home Fleet out of the area while German forces crossed the English channel. Field-marshal Göring, Commander-in-Chief of the *Luftwaffe*, had about 2600 aircraft at his disposal. These comprised both bombers, such as the Heinkel He 111 and Junkers, Ju 87 and Ju 88, and fighters, mainly Messerschmitt Bf 109s and Bf 110s. The date of the air offensive, the famous 'Adlertag' (Day of the Eagle), was fixed for 10 August 1940.

Göring's instructions were precise: first, they were to attack all airfields where RAF fighters, particularly Spitfires and Hurricanes, were based, and put the fighters and their airfields, out of action, and, secondly, they were to paralyse aircraft production by attacking and destroying all aircraft factories.

The German air offensive actually began on 12 August. The attacks took place by day, according to plan, and involved formations of hundreds of aircraft. However, day by day, hour by hour, the British fighters systematically placed themselves in advantageous positions. Taking-off from their airfields, they managed to time their confrontations with the enemy to take place over the English Channel, much to the surprise of their victims, normally the German bomber pilots who did not expect to meet the enemy so soon. 'How do they do it?' they asked themselves. *Luftwaffe* commanders, however, were well informed of the existence of strange and very high antennae located along the south coast of England, and, finally, understood how it was possible for the British to detect the arrival of the invading air formations so soon.

In 1939, the German intelligence services had informed their High Command of the construction of antennae as high as 300 feet along the English coast from Southampton to Newcastle. At first, the real significance of these antennae had escaped the Germans whose undercover agents in England spoke of radio transmission stations operating on very short wavelengths, from 1.5 to 2 m. In fact, the use of these wavelengths was merely a cover for the real wavelength of 40 ms used by the British in their new electronic system of long-range sighting, the Chain Home radar stations.

The explanations furnished by the German secret services did not allay the suspicions of their leaders, particularly Hitler, who also wanted to know what point the British had reached in their preparations for war, above all, in the field of radar. So, on 2 August 1939, just before the outbreak of World War Two, one of the last German airships, *Graf Zeppelin*, took-off from an airfield in northern Germany and headed for the English coast: its mission was to intercept and record the emissions coming from those strange antennae with the aim of analysing their characteristics to find out whether the British had a radar better than that currently being developed in Germany.

Some extremely sensitive receivers and other special electronic measuring devices had been installed on the airship. Several specialised technicians were on board, as well as the head of the *Luftwaffe's* Signal Corps, General Wolfgang Martini.

The airship cruised along the chain of antennae and the technicians on board tried various means to tune in to the British frequencies but no suspicious signals were received. The reasons for the failure of this mission are not known even today. One theory is that the British, having been forewarned of the imminent mission, had detected the airship by radar long before it reached their coast and had immediately ordered their radar stations not to transmit. Another theory is that the receivers on board the airship did not cover the frequency band used by the British, especially the short-wave bands, and therefore could not pick up those transmissions. Others think that the receiver on board the airship had broken down just after take-off and the operator had not had the courage to notify his superiors!

Whatever the reason, the fact remains that, after that flight, the Germans tended to underestimate the danger: Marshal Göring was convinced that they need not worry too much about British radar and, furthermore, that it was not worth spending more time and money on electronic research to develop new radar equipment. According to Göring, the war would be won by the Third Reich in a very short time, thanks to the extraordinary power of the *Luftwaffe* and the *Wehrmacht*.

Consequently, many electronic technicians and engineers were removed from radar research laboratories and employed in other sectors while, in Great Britain, no less than 3000 highly-skilled people were employed to study all aspects of radar, their allotted funds being much higher than the Third Reich's.

The flight of the airship *Graf Zeppelin* is memorable mainly because it was the first mission of electronic intelligence (ELINT), today a routine activity of all modern armed forces.

A few days later, World War Two broke out. The *Luftwaffe* went from success to success in the skies of Poland, Norway and France, where there were no traces of equipment capable of detecting aircraft from afar.

However, when the Battle of Britain began, that invisible electronic wall which had been erected along the English coastline began to annoy Göring. It was obviously making it easy for the RAF to counter the German attacks, so, two or three days after the fighting began, Göring gave orders for it to be attacked and destroyed. The frequency employed by the Chain Home had, by this time, been pin-pointed and, by listening to enemy transmissions, the Germans had determined that the RAF fighters were guided by ground command centres using the new system of sighting whose eyes were, in fact, those strange antennae along the coast.

The first attack was launched on five coastal stations. German fighter-bombers, each carrying two large, 500-kilo externally mounted bombs, made lightning attacks on the antennae, and all five stations were hit and seriously damaged. Despite the fact that only one antenna had actually collapsed, all five had, in fact, been silenced instantaneously.

But three hours after the attack, the British radar stations were again transmitting! Actually, it was a trick devised by the British to make the Germans think that the Chain Home had not been seriously damaged; they had installed some ordinary transmitters to give the impression that the destroyed equipment was still working. In fact, the new apparatus were not capable of receiving echoes, as they were only transmitters, and could not therefore sight any target. The Germans, however, thought that the damage had been repaired already and were convinced that it was useless to attack the antennae since they could be 'silenced' for a couple of hours at most. So the British scheme paid off, as the *Luftwaffe*, under the illusion that the Chain Home was indestructible, thereafter refrained from attacking it during the entire remaining period of the Battle of Britain.

During August 1940, formations of hundreds of German bombers and fighters crossed the English Channel to attack RAF airfields and their hangars. But the RAF fighters, about 700 Spitfires and Hurricanes in all, always managed to be in the most favourable position for intercepting and shooting down the raiding aircraft, particularly bombers. On 26 August, after barely two weeks of combat, the *Luftwaffe* had lost about 600 aircraft, while the RAF had lost only 260. However, RAF Fighter Command was also in dire straits as it did not have enough reserve pilots.

At this point, Hitler intervened with the order to cease bombing enemy airbases and, instead, to systematically bomb London. This change of targeting, while providing a welcome respite for the exhausted British fighters, did not have any really decisive effect. However, it soon became apparent that the German He 111 and Ju 88 bombers were too lightly armed to defend themselves and were thus extremely vulnerable to day-time attacks by Spitfires and Hurricanes. Moreover, even the most advanced *Luftwaffe* fighters, such as the Bf 109, did not have sufficient endurance to operate both as escorts for the slower bombers and as free fighters.

Having been defeated in day-time combat by the omnipresent RAF fighters, the Germans decided to change tactics and commence night bombing. Obviously, adequate navigation and blind bombing systems were required by the attacking forces, while the defenders had to deal with the problem of how to counter this move and defend the country from such attacks.

This was the beginning of a new phase in the air battle of Britain, or rather a new type of war which the British Prime Minister, Winston Churchill, called the 'Wizard War'. He was specifically referring to the electronic countermeasures (ECMs) employed by the British to neutralize the radio navigation aids used by the German aircraft. Churchill wrote:

> This was a secret war, whose battles were lost or won unknown to the public, and only with difficulty comprehended, even now, to those outside the small high scientific circles concerned. Unless British science had proven superior to German, and unless its strange, sinister resources had been effectively brought to bear in the struggle for survival, we might well have been defeated, and defeated, destroyed.[1]

For a better understanding of how this secret war between the Germans and the British developed, we must go back a few years to see how the former acquired their radio-guided bombing technique, used by the *Luftwaffe*'s bombers, and how the latter discovered this system.

In 1930, the German company Lorenz designed and developed a

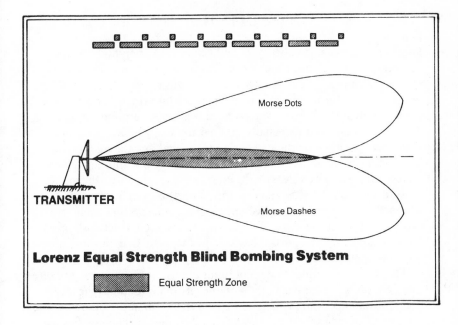

Morse Dots

TRANSMITTER

Morse Dashes

Lorenz Equal Strength Blind Bombing System

Equal Strength Zone

radio-electric navigation system to be used for night landings or for landings in conditions of poor visibility. The Lorenz system marked the flight-path by the equal-strength method, used today in many radio navigation systems. It consisted of two identical directional antennae, placed side by side in such a way that their radiation patterns overlapped. The antennae were connected to two transmitters which were identical except for their modulation: one emitted a series of Morse code dots and the other a series of dashes. A mobile receiving station (e.g. an aircraft) moving within the overlapping section would hear both signals at the same time and, since they were complementary, a continuous signal, or uninterrupted sound, would be heard. This would enable the pilot to know that he was on course. If the aircraft moved off course, the pilot would hear either a series of dots or a series of dashes and could thereby easily deduce which side of the equal strength line he was on. By comparing the relative intensity of the two signals, he could also correct his course to get back on the line leading to the transmitting station (in this case the airbase) which functioned as a directional radio-beacon. The ingenious Lorenz system was immediately adopted in both civil and military airports not only in Germany but also in many other countries, including Great Britain.

In 1933, the German scientist Dr Hans Plendl began to study the possibility of using the Lorenz system to increase the accuracy of bombing systems in conditions of bad visibility or at night. Dr Plendl's system, called *X-Gerät* (X-apparatus), consisted of a certain number of Lorenz beams, one of which was the main radiopath-guide along which the air formation was to navigate, while the others, beamed across the main radiopath, interrupted it at predetermined intervals. These secondary beams usually crossed the main beam above places which were marked on the navigation map, enabling the pilot to know his exact position. This system was combined with a 'time-command' which automatically released the bombs when the main beam was crossed by the final secondary beam. With this system, bombs were dropped on target at night with an accuracy which was truly exceptional for those days. The Germans installed a network of *X-Gerät* systems on the north coasts of France and Belgium just after their occupation.

This so-called 'blind' system of bombing had its baptism by fire on the night of 14 November 1940, with the city of Coventry as the target. Two formations of approximately 450 German bombers took-off in the middle of the night from the airfield of Vannes in occupied France. Nearly all the bombers were equipped with the new Plendl apparatus and, with the guidance of the *X-Gerät* beams, they reached their target and dropped their bombs on the city centre, practically razing it to the ground. That night marked a big step forward in the ever more indiscriminate use of bombers against defenceless civilians, to be followed by the bombing of London, various German cities and, finally, Hiroshima.

This method of night bombing adopted by the Germans was not, however, a complete surprise to the British. On 4 November 1938, the British naval attaché in Oslo had received a secret file from a German citizen claiming to be a 'well-meaning scientist'. These papers revealed that the Germans were constructing a whole range of new, secret weapons, such as missiles, rocket-propelled bombs and magnetic mines, and that they were developing an electronic system using radio beams which would enable an aeroplane to measure its distance from special ground stations. Mention was also made of top secret research which was being carried out on the island of Usedon in the Baltic Sea at a small town named Peenemünde. Most of the things described by the mysterious German scientist were completely new to the British but the little they already knew concerning German weapons corresponded perfectly to what was contained in the file.

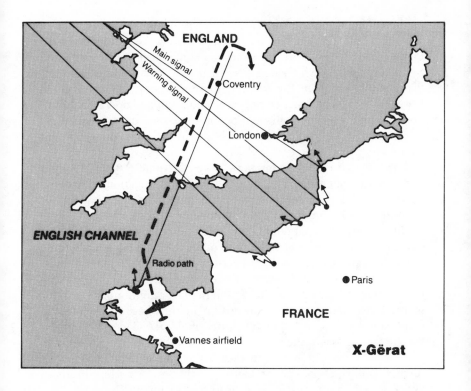

Of course, the document gave rise to strong differences of opinion among the intelligence service, military staff and the scientists who were researching and developing new weapons. Some were convinced that it was a trap designed to mislead the British war effort or a propaganda measure intended to discourage Great Britain from declaring war on Germany; others maintained that the Germans wanted to mislead British scientists and technicians into fruitless fields of research. One group, however, felt it wise to investigate the details of this precious piece of information closely: among these was Winston Churchill who, when the war broke out and the threat of electronically-guided bombing was imminent, wrote: 'If these facts correspond to the truth, they represent a deadly danger.'

He immediately set up a committee of scientists to study, not only what was alluded to in the Oslo papers, but also the possibilities of using applied electronics for military purposes, a use which Churchill was advocating more and more strongly.

Meanwhile, the Germans had decided to perfect their electronic method of bombing as it had shown two serious shortcomings. First of all, with this system the bombers had to fly along the radiopath for too long a period of time with the almost certain danger of being attacked by British fighters which always seemed to be positioned along their flight paths, much to the surprise of the Germans. Secondly, the Lorenz system was rather complicated and pilots and operators had to undergo long training courses. For these reasons, the Germans had started studying simpler, improved systems of radio-guidance and were soon trying them out.

The fear that the Germans were using electronic systems for night bombing gained ground in Britain following interrogation of captured *Luftwaffe* pilots and analysis of electronic equipment found in the wreckage of a German bomber shot down on British territory.

On 21 June 1940, all doubts about the existence of such electronic systems disappeared when the pilot of an RAF Anson on a routine electronic reconnaissance flight heard in his earphones something he had never heard before: a series of particularly clear and distinct dots transmitted in Morse Code shortly followed by a continuous signal (or whistle). Still flying on the same course, he then began to hear in his earphones a series of dashes trasmitted in Morse Code. The Anson was, in fact, crossing the radio-beam emitted by a German station to guide the German bombers to their target. This incident provided further confirmation of the truth of what was written in those mysterious documents handed over to the British attaché in Oslo.

Following this lucky discovery, the British began to study all the possible ways of opposing the German system in order to reduce or, if possible, neutralise its effectiveness. These became known as electronic countermeasures (ECM).

One countermeasure which was considered by British scientists was to transmit a continuous noise, produced by an electric machine, on the same frequency as that used by the Lorenz system. A medical cauterising instrument turned out to be the most suitable machine for this task and the larger London hospitals were immediately consulted! The electrical discharges produced by such an instrument would have disturbed the German transmissions enough to render their guidance-system useless. Another way of obtaining the same result would have been to place a microphone near the spinning propeller of an aeroplane and transmit the noise on the same frequency as that used by the Lorenz system (200–900 Kc.)

However, these ways of interfering electronically with the Lorenz system had the serious disadvantage that the enemy would be aware that the interference was deliberate and, consequently, that their Lorenz system had been discovered. They would then think up some devilish new method which, no doubt, would have even more serious consequences for the British cities which had by now become the main target of the Third Reich's night bombing.

To avoid this, British scientists devised a means of deceiving German pilots by transmitting signals of the same type as those they expected to hear but containing deliberate distortions (such as direction of arrival) which were intended to deceive them without arousing any suspicion. This system had to be put into effect immediately as the Germans had already inaugurated their radio-guidance system in the bombing of Coventry and were relentlessly returning every night to bomb the island.

Following a period of intensive research, the British finally found an antidote to the Lorenz system, which they called Meacon (Masking Beacon). This countermeasure consisted of retransmitting the signal emitted by the Lorenz after having tampered with it. A receiver and a

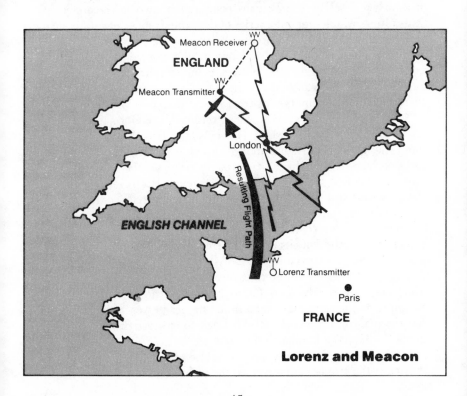

Lorenz and Meacon

transmitter were installed about 10 miles apart in the South of England: the receiver picked up the Lorenz signals and sent them by cable to the transmitter which then immediately retransmitted them using a much greater power and a directional antenna which emitted the radiobeam in a slightly different direction from that of the original beam coming from the Lorenz apparatus. At a certain point, German pilots flying along the radio beam would hear two signals, the original becoming fainter and the retransmitted one becoming stronger. They would automatically pay attention to the stronger signal which would take them off course away from the target to be bombed.

With this trick, German pilots ended up bombing open country instead of the assigned cities, and, in many cases, lost their way and had no alternative but to land in England. After some time, the Germans realised that their Lorenz system had been completely neutralised by British countermeasures and immediately modified their radio-guidance system for navigation and bombardment.

The new system (named 'Knickebein' by the Germans and referred to as 'Headache' by the British) consisted of two inter-connected transmitters which sent out a series of dots and dashes. The difference between this and the old Lorenz system was that, instead of having many cross beams, there was only one, which crossed the main beam exactly above the target city. Besides being simpler, the new system was also much more accurate as the continuous signal was within a 3 degree sector with a margin of error of less than one kilometre.

As soon as this new system was introduced, the German bombers began to achieve better results. However, the British had found out about the 'Knickebein' some months previously when they found, in the wreckage of a Heinkel 111, a paper headed 'Navigational Aid' which mentioned 'Knickebein' and contained data regarding times, places, routes and so forth. Interrogation of captured German pilots and careful examination of every radio apparatus found in shot-down German bombers soon revealed the main characteristics of the 'Knickebein' (especially its operating frequencies, the most important of which was 30 Mc/s) and the British promptly came up with the 'Aspirin' (electronic countermeasure) for the 'Headache'. They reinforced one of the two German signals (dots or dashes) by transmitting the same signal at a much higher power with the result that the main beam was slightly inclined to either right or left, thus taking the German bombers off course.

The British Radio Intercept Systems also managed to pinpoint over which city the cross-beam traversed the main beam in time to alert the population and organise air defence, concentrating RAF fighters in the area, of the expected attack. At this point, heavy losses had been suffered by both sides. By the end of September 1940, the Germans had lost 1100 aircraft and the British at least 650 fighters.

By now, it was clear that the German plan to conquer the skies above the Channel and the south of England had failed and Hitler had to put off indefinitely his much-desired invasion, Operation 'Sea Lion'. Moreover, in the late autumn of 1940, bad weather forced the Germans to slacken their pace and cut down on air raids which, by now, were nearly always at night as only the darkness could protect the bombers from the inevitable relentless attacks by British fighters.

Meanwhile, an intense struggle was going on in the laboratories of both countries in an attempt to devise more sophisticated electronic equipment, especially now that radar was proving more and more to be an indispensable means of pinpointing the enemy and directing fire against him. However, as soon as one side came up with a new electronic measure, an appropriate countermeasure was immediately devised by the other side to neutralize or at least reduce its effectiveness.

During this dramatic phase of the war, the BBC had been ordered to use the same frequency for all its transmissions as it had been discovered that German pilots who had lost their way, due to British ECM or bad weather, used British Broadcasting Corporation (BBC) radio stations to get back on course. They used the direction-finders they had on board to measure the course or bearing to two or three BBC radio stations in order to fix their position, by triangulation.

Another radio station which was used for military purposes was Radio Paris. Unlike the BBC, which mainly transmitted entertainment, war news and political speeches designed to keep up morale, Radio Paris offered a continuous, round-the-clock programme of lighter fare in the form of songs and variety shows, interspersed by Nazi propaganda. These programmes were listened to by a great number of British people who tolerated the Nazi propaganda in the same way we tolerate the commercial breaks which interrupt our television programmes nowadays.

After a while British listeners noticed that the volume would increase every now and then and they would have to turn their sets down; they also noticed that this usually happened just before a German air raid. This strange coincidence was soon made known to the proper authorities who, after detailed investigations, discovered that, indeed, the transmissions increased in volume in those cities which were bombed shortly after and also that the volume was proportionately weaker with distance. It was concluded that the Germans must be using Radio Paris to guide their bombers over the cities.

This was, in fact, the case; before every air raid, the Radio Paris transmissions were switched from the normal omnidirectional antenna to a highly-directional antenna aimed at the city to be bombed. The German pilots would thus be directed to London or Liverpool simply by listening to the French songs transmitted by Radio Paris! Another narrow ray, intersecting the main beam above the target city would signal the bomb-release point.

This new system, which operated on a frequency of 70 Mc/s and was called 'Ruffian' by the British, is something of a mystery even today. It is difficult to understand how the Germans managed to transmit such very narrow (3 degree) electromagnetic beams with the limited electronic technology then at their disposal.

The British took a long time to detect this fiendish system, but

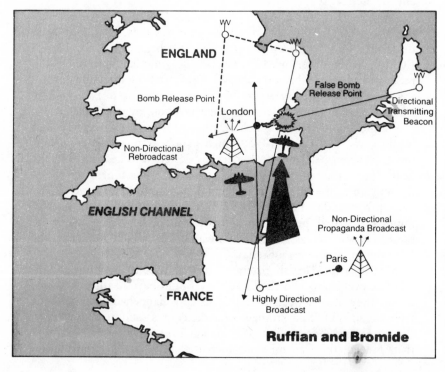

Ruffian and Bromide

eventually came up with an effective countermeasure which they called 'Bromide'. This consisted in re-transmitting the Radio Paris programmes on the same frequency but using an omnidirectional antenna thus neutralizing the German's directional aid.

With this electronic countermeasure the British managed to completely disorientate the German bombers which flew haphazardly over Britain dropping their bombs at random. Later, the British managed to achieve directional transmissions which enabled them to induce the bombers to drop their bombs at sea. To keep the Germans in the dark regarding the success of their electronic countermeasures, the British Press attributed these random bombings to diversive actions organized by the Germans against the Spitfire bases. This countermeasure did not last long, however; at the beginning of 1941, the Germans came up with another bombing-aid system which they named 'Benito' in homage to their Italian ally, the 'Duce' of Fascism.

In this period, frequency modulation was almost unknown and so the Germans, convinced that the British had no means of listening to

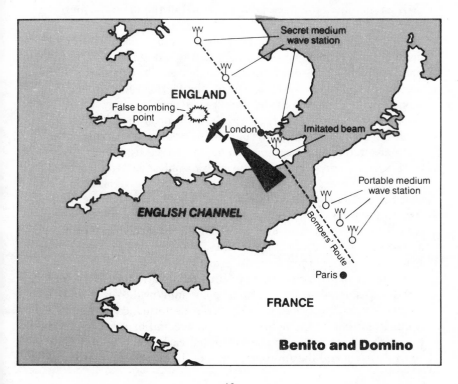

this type of transmission, set about trying to exploit it in order to elude British surveillance. A number of secret agents, equipped with portable FM (frequency modulation) radio sets, were positioned along the main routes in Britain and France and could furnish German pilots with positional and other information, including their exact distance from the target.

It was not easy for the British intelligence service to understand what was happening but they eventually managed to intercept the communications between the German secret agents and pilots and immediately devised a simple but effective countermeasure. They employed German-speaking operators who would, using the same frequency, transmit false information to the enemy pilots. This countermeasure, called 'Domino', was so effective that several German pilots were induced to land at British airbases without realizing it!

However, 'Domino' was not without its drawbacks. One grave consequence of its shortcomings was when, on the night of 30/31 May 1941, the 'Domino' operators accidentally directed a formation of German bombers to attack Dublin, capital of the neutral country of Eire.

Finally, the Germans resorted to fire-bombing; the glow from the fires caused by these bombs was sufficient to illuminate the area to be bombed by the successive formations of planes. The British reacted by setting up huge fires to act as bait for the German formations. This was, of course, done outside London in open country where the Germans, ignorant of the trick, regularly dropped their bombs.

However, by now, the Battle of Britain was already petering out and German aircraft were beginning to be transferred from France to the eastern front in preparation for the invasion of Russia. After months of fierce air combat, heavy bombing and desperate fighting against Britain's air defences, the Germans had failed to conquer the skies of Britain and their plan to invade the island had gone up in smoke. The RAF had emerged victorious even though its losses were probably almost as heavy as those of the enemy. Approximate losses were 1500 British and at least 1700 German fighters.

There are many factors that favoured the final victory of the RAF, rightly considered to be a turning point in World War Two. The victory is usually credited to the superior performance of the British Spitfire and Hurricane fighters, the courage and experience of their pilots, the effectiveness of the integrated early warning and fighter control system and the appropriateness of the tactics of No 11 Group,

Fighter Command; moreover, the *Luftwaffe* made a number of tactical errors. However, a careful study and analysis brings to light other considerations which documentation and statistics available today confirm.

The British were on home ground and, whilst their fighters were obviously faster than the slow German bombers, the latter had to endure long flights, often in bad weather conditions over the threatening waters of the English Channel and the North Sea. The British also had the benefit of an excellent warning and command system, the very effective electronic countermeasures, which deceived the German pilots by leading them to believe that they were attacking the right target when, in fact, they were often dropping their bombs in open countryside or in the sea. It has been estimated that only a quarter of the bombs carried by the Germans reached the urban areas and factories which were their targets.

Secondly, those electronic countermeasures devised to interfere with navigation caused the German pilots to fly in an insecure state of mind, not knowing whether to rely on their navigational instruments or on their senses, which had a debilitating effect on the operative efficiency both of the crew and of the aircraft.

Despite alternating periods of success and failure, the various systems of electronic countermeasures devised to neutralise or reduce the efficiency of the radio navigation systems used by the enemy carried great weight in deciding the final outcome of the Battle of Britain.

6

Electronic Battles in the Atlantic

Another important chapter in the history of electronic warfare was the desperate struggle between Axis submarines and Allied air and sea anti-submarine forces in the so-called Battle of the Atlantic.

At the beginning of the war, the only means available for detecting submarines was Asdic (Anti-Submarine Detection Investigation Committee), now commonly referred to as Sonar (Sound Navigation and Range). It involves sending out sound waves into the water which, on meeting an object, are reflected back; the distance of the object is calculated by measuring the time taken by the wave to return. This is called echo-ranging.

However, in the summer of 1940, Admiral Dönitz, Commander-in-Chief of the German submarine fleet, decided to make a radical change in the tactics of submarine warfare. He had noticed that the Allied convoy's naval escorts mainly consisted of rather old destroyers since the best ships of the Royal Navy were being used to fight German merchant raiders. Taking advantage of their weak defence, Dönitz decided to attack the enemy convoys by night on the surface rather than under water. In these conditions, Asdic (whose range was minimal near the surface) would be impotent against the fast German U-boats which, protected by the darkness, would be able to attack and retreat without submerging. At night, the low superstructures of the submarines would be difficult to make out in the vast expanse of the ocean while the massive dark forms of the merchant ships, standing out against the slightly lighter sky, would be easy targets for the submarines.

The commanding officers of the U-boats, realising their advantage, became more and more audacious in their attacks, penetrating into the midst of the slow convoys and causing enormous damage. They were also greatly helped by the German radio interception service, Service B, which picked up and deciphered not only the messages transmitted by the British convoys at sea but also the route instructions transmitted by the British Admiralty to the ships.

As the number of merchant ships sunk rose day by day, Britain began to face the awful prospect of having her supply lines from the British Empire and the United States cut off completely. Alarmed by this prospect, the British decided to install on some of their escort ships and RAF Coastal Command aircraft ASV (Anti-Surface Vessel) radar. However, the mark I performed poorly in submarine warfare and, at the beginning of 1941, was replaced by the Mark II, which was installed on aircraft. With the aid of this apparatus, an aircraft flying at an altitude of 1500–3000 feet could detect a submarine on the surface at a distance of 8 miles. The Mark II also proved inadequate as, when the aircraft closed in to bomb the submarine, sea-clutter (echoes reflected from the sea) masked the target echo on the radar screen making night bombing of submarines on the surface ineffectual.

The enemy, however, was unaware of this limitation and so the presence of radar on board the aircraft helped to lower the number of British merchant ships sunk, at least in the coastal waters off the west coast of the British Isles out to the limits of the range of Coastal Command aircraft.

At this point, the German submarines began to use a new tactic introduced by Admiral Dönitz. This was the 'wolf-pack' method of attack which involved concentrating a number of submarines at strategic points so that, when an enemy convoy tried to pass, it was attacked from all sides for several days running. This new tactic, applied mainly out of range of British aircraft, created great difficulties for the naval escort units and caused havoc in the convoys; many British merchant ships were sunk as a result of this new method of attack. After the United States entered the war, German submarines could also operate along the American coastal routes used by a great number of merchant ships which, unarmed and unescorted, were totally defenceless against the underwater menace. Thus, the number of ships sunk rose astronomically: in the months of May and June 1942 alone, 200 merchants ships were sunk along the American coasts!

While this fierce combat was going on at sea, another challenge was being prepared in the laboratories. Scientists were busy devising a series of electronic measures, countermeasures and counter countermeasures which were to sway the course of the Battle of the Atlantic.

The Allies began by installing a new L-band radar, i.e. operating on frequencies between 1000 and 2000 Mc/s, in aircraft which had the range and endurance to allow all the main convoy routes between Great Britain and the United States, to be covered from bases in these nations. Thus, in the summer of 1942, Allied aircraft were able to begin night bomb attacks on German submarines using a very powerful searchlight—the Leigh Light—which could illuminate them from a distance of about a mile. Now that the Allies had found a way of covering the whole Atlantic and had overcome the problem of loss of radar contact at close quarters by using a searchlight, the number of German submarines sunk began to rise.

The Germans countered this by installing Metox Radar Warning Receivers (RWR) on their submarines. As previously mentioned, this type of receiver was able to pick up the enemy signal before it was itself detected by the enemy. The French firm, Metox, already had the RWR in stock but the antennae had to be hurriedly improvised by winding wire round a wooden cross; this improvised antenna was jokingly referred to as the 'Biscay Cross', alluding to the Bay of Biscay where the German submarines had to contend not only with the storms for which it is famous, but also with concentrations of Allied aircraft and warships. The Metox, by providing early warning of the enemy aircraft or ship, enabled the submarine to crashdive in time.

This countermeasure had immediate effect and the number of submarines sunk declined noticeably. The Allies, of course, realised that something new was happening in the electronic field and started work on a new radar, the Mark III. This radar worked on a frequency of 3000 Mc/s corresponding to a wavelength of barely 10 cms, in the S-band (2000–4000 Mc/s), a much higher frequency than its L-band predecessor. The Mark III was introduced in the early months of 1943. The Metox receivers, which were supposed to warn the German submarines of approaching enemy aircraft or warships, unable to intercept high frequency transmissions, remained silent. Thus the U-boats, relying on their Metox receivers, having unsuspectingly surfaced to recharge their batteries were easy targets for the Allied aircraft equipped with the new radar.

As submarine losses began to increase once more, German technicians frantically tried to discover what had changed in the Allies' locating methods. Although the surviving U-boat commanders reported that no emission of electromagnetic energy had been picked up by their receivers before the attack, for some reason the German technicians did not consider the possibility that a higher frequency was being used. Instead, they surmised that a new device was being employed using infrared rays, for example, an extremely sensitive radiometer capable of detecting the heat emitted by a U-boat's engines. So, following this false trail, they embarked on a lengthy project to suppress the heat emitted by the submarines. After months of research and experiment, heat-shields were installed on the sides of the U-boats but the only effect this had was to reduce their speed. Meanwhile, the number of U-boats sunk was rising; in the months of May and June 1943 alone, about one hundred were lost.

The *Luftwaffe* came to the aid of navy technicians when they found, among the wreckage of a British aircraft shot down near Rotterdam, some pieces of the radar H₂S which revealed a technology hitherto unknown to German experts. Through this lucky find, they came to realize that the Allies had invented the famous Magnetron, a very sophisticated electronic tube operating on a wavelength of about 10 cms. German industry immediately set about building an RWR which could pick up transmissions in the S-band. The new receiver, which was called 'Naxos', took time to develop and proved to be inadequate, having insufficient sensitivity and a range of only 4–5 miles.

Meanwhile, more and more U-boats were being sunk by the Allies and so the Germans tried other methods of avoiding detection. One such method employed decoys code-named 'Bold', in the form of rubber balloons launched from the submarine which rose to a height of about 30 feet. These were attached to one or two metal cables intended to reflect the enemy radar emissions and thus create a false echo. The balloons were moored to floating buoys which used to drift about but the chances of deceiving patrolling aircraft were very slim indeed.

Towards the end of 1943, a measure of success was obtained with the *Schnorkel*, a tube fitted with a special valve which enabled the submarines to recharge their batteries while submerged. These were covered with a special anti-radar material which absorbed instead of reflected enemy radar emissions.

When the long-awaited 'Naxos' receivers were finally ready, it was too late; too many submarines had been sunk and the battle of the Atlantic was, by this time, irretrievably lost.

Interception of U-boat radio transmissions contributed greatly to the Allies' victory in the battle of the Atlantic, especially in the period of greatest losses of convoy ships. The submarines would periodically surface, usually at night, to recharge their batteries, check their position and transmit operational reports to headquarters or exchange information by radio with other submarines in the area. These transmissions were intercepted by Allied escort ships which had direction-finders on board and could, by taking bearings, locate the enemy vessel. Its position would then be communicated to the hunter-killer groups which had the task of finding and sinking the submarine. These groups, which usually consisted of two or three destroyers or frigates, would head for the point indicated and mercilessly hunt down the unfortunate enemy.

To avoid this, the Germans devised a system for making extremely, rapid or 'squirt' transmissions; they recorded the messages and

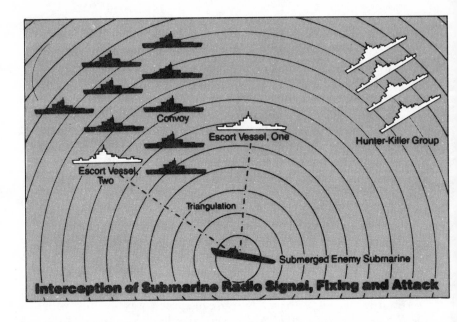

Interception of Submarine Radio Signal, Fixing and Attack

compressed them by speeding-up the recorder until the message could be sent in less than one second. The recorder would automatically slow down the recording for normal reading. The direction-finders then in existence were not fast enough to intercept and locate the transmitter in such a short time and so the U-boats were able to pass fairly peaceful nights for a while.

But, in 1943, the Allies came up with a countermeasure, an automatic direction-finder called 'Huff-Duff' which was able to pick up the brief transmissions and calculate the direction in the fraction of a second that they lasted. 'Huff-Duff' sets were installed not only on board ships but also at shore stations favourably located to get a good triangulation of the transmissions intercepted. As soon as a German submarine transmitted a message, the ground stations and ships at sea could immediately find its position and send anti-submarine ships and aircraft to attack and sink it.

The Battle of the Atlantic constitutes an important lesson for those in charge of the planning and conduct of electronic warfare: it teaches that it is not sufficient to know what kind of equipment the enemy is using on the battlefield at the moment but that it is vital to find out what is being developed for use in future operations. General Martini had made a wise decision when, in 1939, he decided to fly along the

British coasts in the airship *Graf Zeppelin*. He thus effected the first electronic reconnaissance flight, an operation aimed at finding out what the enemy is doing in the field of electronic warfare. The *Luftwaffe* should have carried on with such activity during the war by sending not just bombers and fighters over Britain but also a few aircraft equipped to intercept electromagnetic emissions present in the British sky. The German electronic industry already had sufficient know-how to explore the electromagnetic spectrum; crystal video receivers, ideal for this task, were already in use and the Germans could, and should, have employed them to pick up the pulses transmitted by the new British radar sets while these were still being tested.

In neglecting electronic reconnaissance, the German high command not only underestimated the impending threat but also deprived themselves of the possibility of acquiring knowledge of technical innovations which might have proved extremely useful for developing electronic countermeasures capable of neutralising these impending threats.

7

The 'Channel Dash'

In the early dawn of 22 March 1941, the German battle cruisers *Scharnhorst* and *Gneisenau* sailed into the naval base of Brest in occupied France. Both ships had completed a long mission in the Atlantic and were in need of repairs after having sunk over twenty British merchant ships. Two months later, another German warship, the heavy cruiser *Prinz Eugen*, took refuge in the same port. A few days earlier, she had fought against the British Fleet with the battleship *Bismarck*. The *Bismarck* had been sunk while the *Prinz Eugen* had escaped.

During the repair period, the three big German ships, although well camouflaged, were detected by the RAF who immediately began to bomb the port night and day. The ships were damaged more than once during the bombing raids and, after a while, the German high command decided to remove them to a safer port in Germany. On the opposite shore, the British realised that the enemy ships would sooner or later try to leave Brest to get away from the daily raids, and began to make arrangements to prevent them doing so.

Transferring the three ships from Brest to Germany would be extremely risky. Hitler himself had compared the situation to that of a patient dying of cancer; not operating meant death, while operating, however risky, offered some chance of salvation. So, it was decided to operate. The first thing to do was to decide which route the ships would take to reach Germany. There were two possibilities, one worse than the other. The first was to sail west and north around the British Isles, and then through the North Sea; but, as the sinking of the *Bismarck* had shown, the British Fleet would have plenty of time to intercept the ships and sink them. The other possibility was to sail through the English Channel; this route had the advantage of being much shorter but involved passing right under the noses of the British with consequences which can easily be imagined. However, the second route offered some chance of avoiding attack by the heavier warships of the Home Fleet as the British, ironically, had transferred most of their ships to northern ports to avoid the *Luftwaffe* raids. The main dangers of the English Channel were the presence of large numbers of torpedo boats and aircraft, long-range guns positioned along the coasts of the Straits of Dover and, of course, floating mines. It would be essential for the German ships to reach the Straits of Dover without being discovered if the transit was not to end in total disaster. If they managed to reach the Straits of Dover without being sighted, simple mathematics showed that it would be impossible for the British Fleet to reach and attack them thereafter.

Having weighed the pros and cons, the German naval command chose the shorter route through the English Channel. They were then faced with the dilemma of whether to leave Brest by day or by night. Departure by day meant that they could pass through the Straits of Dover at night while departure from Brest by night would leave them exposed to day-time gunfire in the Straits. Eventually, they decided to leave Brest by night. This decision was based on the fact that their fear of British aerial reconnaissance was greater than their fear of gunfire.

The head of the *Luftwaffe* Signal Corps, General Martini, had been dealing personally with enemy radar since August 1939 when he tried to intercept, without success, electromagnetic emissions in the skies over Britain (See Chapter 5). With the fall of France in 1940, receivers had been positioned along the northern coast of France to intercept British radar transmissions. In this way, the Germans had discovered the main charactertistics of many British radars (frequency, duration of pulses, etc) and their locations. General Martini was put in charge of radar jamming for the 'Channel dash' with the aim of delaying as far as possible British detection of the three German warships.

Although radar is an effective means of reconnaissance and gun-laying, being able to penetrate fog and darkness, it is susceptible to jamming and deception. This vulnerability of radar is a main factor in electronic warfare. A radar set is vulnerable to jamming because the echoes received from targets are generally very weak. When a radar receiver is made sensitive enough to pick up these weak target echoes, it can easily be overwhelmed by signals from a more powerful jammer transmitting on the same frequency and aimed directly at it. To jam the British radars covering the English Channel successfully, it was obviously essential to know their exact frequencies and their approximate geographical locations, information well known to General Martini.

German industry immediately designed and manufactured special jammers capable of saturating the British radar receivers and blanking their cathode ray tubes. Martini had these installed in the vicinity of Ostend, Boulogne, Dieppe, Cherbourg, Rotterdam and other suitable locations along the northern coast of France. Each jammer was assigned a victim from among the British radar stations.

General Martini's simple but ingenious plan was aimed at preventing the British from using their radar without their realizing that they were being deliberately jammed by the enemy. Two months before the date fixed for the departure of the ships from Brest, systematic jamming of British radar was initiated, in the hope that the British would think it was due to atmospheric disturbance. At first, the interference lasted only a few minutes but was gradually increased day by day so that the British would get used to it and think that it was due to the particular atmospheric features of the area and, therefore, inevitable. After a month or so the desired result was obtained.

The Germans, with their usual thoroughness, took all sorts of steps to ensure that the departure of the ships would take the enemy by surprise. First of all, only the commanders of the three warships were informed of the plan. Secondly, they set out to deceive the inhabitants of Brest, among whom were a considerable number of British spies and members of the French Resistance, by organising a fancy-dress ball and a hunt to give the impression that they had no intention of moving for some time. Finally, to create the impression that the next mission of the *Scharnhorst*, *Gneisenau* and *Prinze Eugen* would be to attack enemy convoys along the African coasts in the south Atlantic, colonial helmets and barrels of oil marked 'for use in the Tropics' were loaded onto the ships; furthermore, normal dealings with post offices, food suppliers and laundries were kept up, all with the aim of allaying enemy suspicions of an imminent departure.

In spite of all these precautions, the British Admiralty, after carefully weighing up the situation, nevertheless arrived at the conclusion that the three ships were preparing to leave the port of Brest and that they had almost certainly chosen the route via the English Channel. The British planned operation 'Fuller' to prevent their passage to Germany. British reconnaissance flights made on 29 and 31 January had, in fact, revealed that numerous torpedo boats, light destroyers and mine-sweepers had arrived at Brest, a sure indication that an imminent departure was planned. So, on 3 February, the British Admiralty gave the order to put into effect the various provisions which had been made some time ago for such an eventuality. Mines were laid along the route which the ships would probably take, while coastal radar and RAF Coastal Command flights were put on maximum alert. Both contenders were now ready for action, having prepared their strategies down to the smallest detail and in the utmost secrecy.

The departure of the German ships was scheduled for midnight on

11 February 1942. The time and date were chosen to take advantage of the darkness caused by the new moon and the high tide which gave an additional 16 feet of draught, as well as a favourable current of 3 knots. A fake air raid was organised to take place just before midnight; a few bombs were dropped on some deserted sand-banks of the port and the population of Brest took refuge in air raid shelters. While they were waiting for the all-clear, the three ships, escorted by eight destroyers and sixteen torpedo boats, weighed anchor and slowly glided out of the port.

Once out at sea, the crews were finally informed that they were returning to Germany. This news caused great excitement both at the idea of going home and of taking part in such a daring operation. The night was dark and foggy and the ships sailed as close as possible to the French coast along narrow lanes which had been cleared by German mine-sweepers a few hours beforehand. Strict radio and radar silence was maintained by the entire formation, and to communicate with one another the ships used special infrared signalling lamps invisible to the British.

The German naval units had formidable firepower: *Scharnhorst* and *Gneisenau* each mounted nine 280 mm guns, twelve 152 mm guns, fourteen 105 mm guns, and numerous 37 mm machine-guns, and thirty-six launching-tubes for 533 mm torpedoes. The *Prinz Eugen* had eight 203 mm guns, twelve 105 mm guns, twelve anti-aircraft 37 mm machine-guns and twelve launching-tubes for 533 mm torpedoes. There was also the firepower of the destroyers and torpedo-boats. The whole formation was covered by an air umbrella of 250 long-range, fighters under the command of the famous Luftwaffe fighter pilot Adolf Galland, who then had 94 air combat victories and, at the end of the war, was credited with a total of 103.

On the British side, there was the entire Home Fleet and several hundred aircraft of various types, but perhaps the greatest threat was the British radar network along the south coast of England. The Germans dosed their jamming very carefully so as not to arouse the slightest suspicion.

All night long the formation steamed at full speed towards the Straits of Dover without being sighted by British reconnaissance aircraft. As dawn approached, tension grew among the German sailors who expected to be attacked at any moment by the enemy fleet. In the early hours of the morning two Heinkel He 111 aircraft equipped with jammers began their carefully prepared jamming of coastal radar stations; they flew along the English Channel parallel to the south

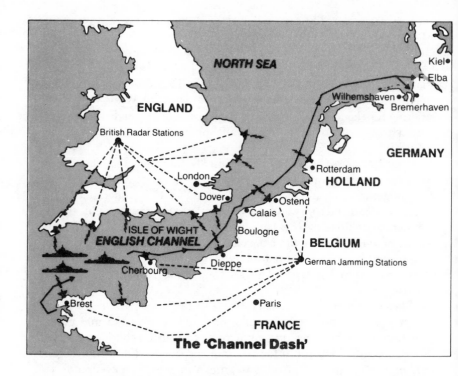

The 'Channel Dash'

coast of England, preventing British radars from detecting the presence of the large air formations which were escorting the German naval units. The ground-based jammers, on the other hand, maintained strict silence until 09.00 hours in the morning; their task was to mask the presence of the German ships and they were to operate only when the formation was approaching the limits of the range of the British radar stations surveying the Straits of Dover. On schedule, these jammers were put into operation and, being perfectly tuned to the frequencies of the British radars, worked so well that some British radars were forced to shut down while others changed frequency in a vain attempt to evade the jamming. One completely unknown radar station started to transmit but this too was immediately blinded by German jamming. In short, the jamming operation was such a huge success that the British operators did not even suspect anything out of the ordinary!

This was a real battle of electronic warfare, the first in naval history. The German formation had now been sailing for about ten hours and was rapidly approaching the Straits of Dover. There was no indication that they had been detected by the British and it seemed as though the brilliant German electronic planning was going to succeed. However, at about 10.00, a British radar started transmitting on such a high frequency that the Germans were totally unable to jam it; it was this new radar station that signalled the presence of enemy aircraft flying low along the French coast. At about 10.45, some patrolling RAF aircraft encountered a large formation of *Luftwaffe* aircraft and, to make their escape, dived low to skim the waves where they finally caught sight of the German ships. Due to an incomprehensible series of delays, British air and naval commands were not informed of the sighting until almost an hour later, at 11.30. It was almost midday when the German formation, passing abeam Boulogne, came within range of British shore batteries covering the Straits of Dover, which opened fire at a range of 29 000 yards. Not one round hit the German ships, however, as the thick fog made it impossible for the British gunners to see where they were firing and they had to rely solely on their radar which did not afford sufficient accuracy. The German shore-batteries installed on the French coast immediately returned fire and silenced the enemy.

By now the battle had begun and the German sailors, although exhausted by their sleepless night, were on the alert, expecting to be attacked at any moment by the British Fleet. It was not long before six Fairey Swordfish torpedo-aircraft of No 825 Sqn, Royal Navy, came in to attack, escorted by five fighter squadrons. However, co-ordination between the formations was impossible as the Swordfish were slow (145 mph) and had to attack at wave-top height. They had to attack without cover and were consequently shot down inexorably one by one, and no torpedo found a target: it was a suicidal attack launched with great courage and the leader, Lieutenant Commander Eugene Esmonde RN, was awarded a posthumous Victoria Cross.

The next attack came from a flotilla of torpedo-boats hastily sent from Dover. These small vessels, more suitable for night fighting, were no match for the German destroyer escorts but, nevertheless, managed to get near enough to launch their torpedoes. The Germans employed smoke-screens and evasive tactics to avoid them and, with their enormous firepower, forced the enemy to withdraw. By now, the Germans were extremely nervous, wondering what the British navy and air force had in store for them. Soon, twelve British destroyers arrived, while 240 bombers took-off from various airfields in Great Britain and a squadron of mine-laying aircraft dropped magnetic mines along the route the Germans ships would probably have to follow. The British destroyers bravely attacked the large German ships, coming up close to launch their torpedoes, but their efforts and courage proved useless and one of their units was seriously damaged by German fire.

In the afternoon, a sudden explosion shook the *Scharnhorst*. The lights went out and the engines stopped. She had hit a mine. The crew silently watched the *Gneisenau* and the *Prinz Eugen* sail by; orders were to proceed at all costs and, if one of the ships was hit or sunk, the others were not to stop and help her.

While *Scharnhorst's* crew was trying to repair the damage to their ship, a desperate air battle was going on in the skies above the English Channel. Thirty-six Bristol Beaufort torpedo-aircraft attacked the German naval formation but their co-ordination was so bad that the attack failed. Meanwhile, *Scharnhorst's* crew had managed to improvise repairs and the warship was proceeding on course.

During the late afternoon, 240 RAF bombers made repeated wave attacks on the German ships but the poor visibility made flying operations difficult, and, in fact, only forty planes managed to press home their attacks while another fifteen were shot down and twenty damaged.

At about 19.00 *Gneisenau* hit a mine but the damage was not serious and she was able to proceed at a speed of 25 knots. To make things worse, a violent storm had broken out in the area and the German ships lost contact with one another and were unable to distinguish the markers left by their mine-sweepers. Then, *Scharnhorst* hit another mine, this time with more serious consequences. She took in about 1000 tons of water and, with many parts of the ship flooded and her steering out of action, she came to a stop and began to drift towards mine-fields and sand-banks.

During the night, the RAF stepped up their operations, carrying out over 740 attack sorties against the German ships, to which the latter replied with anti-aircraft fire—the German sailors had to continually pour buckets of water over the gun barrels which had become incandescent from repeated firing! However, these attacks did not have conclusive results.

Just before dawn, the approach of unknown ships was signalled in the German ships, causing great apprehension as the German formation was in no condition to meet the British Fleet at the particular moment. However, the unknown ships turned out to be two of the German escort ships which had lost contact with the others because of the darkness. In spite of everything, the German formation

Interception equipment used during World War One, operating on a wavelength between 2500 and 25,000 m.

Low frequency amplifer used at telephone interception posts during World War One.

The German pocket battleship *Graf Spee* burning in the estuary of the River Plate.
The inset shows the radar antenna arrowed in the main illustration.

The first Italian radar, called 'Gufo', installed at the top of the forward superstructure of the battleship *Vittorio Veneto*. (Historical Office of the Italian Navy)

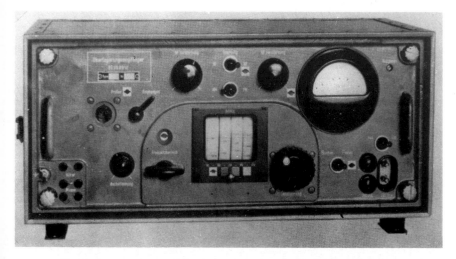

The German 'Metox' interceptor receiver, Fu.Mb.1 (Funkmess Marine Beobachtung). This pulse detector, installed on German ships and U-boats for early interception of hostile radar emissions from RAF Coastal Command aircraft and Royal Navy warships proved to be very useful. The operator could monitor radar emissions of radars operating in the band 113 to 500 Mc/s.

British listening station for interception of radio communications by German ships and U-boats. (By courtesy of *Defense Electronics*)

British Chain Home radar antenna on the Straits of Dover.

'Carpet' jammer installed on Allied aircraft during World War Two.

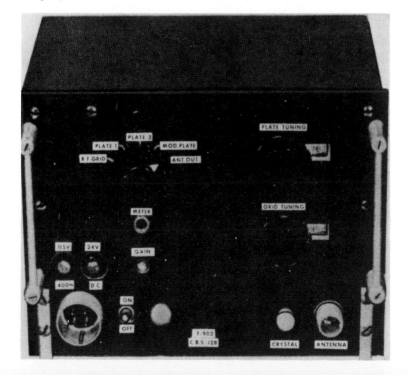

Left: a radarscope showing many target echoes.

Right: the same radarscope being jammed.

The German aircraft support ship *Togo* (upper right), with the ship's early warning 'Freya' (upper left), its 'Würzburg' radar for fighter plotting and anti-aircraft direction (lower left), and the two antennae of the 'Heinrich' jammer (lower right) used to confuse the British 'Gee' navigational guidance system.

German Junkers Ju 88G-1, used mostly as a night fighter, with large 'Flensburg' antenna installed in the nose to intercept radar emissions and establish the direction of incoming RAF bombers. It was tuned to the 'Monica' radar installed in the tails of RAF bombers, enabling the 'Flensburg' to guide the Ju 88 onto them, even from considerable ranges.

The electronic instrument panel of a Ju 88G-1 night fighter. On the right are the two SN2 radar indicators, **A**, and the 'Flensburg' intercept indicator, **B**.

An RAF Avro Lancaster ejecting 'Window', seen from another Lancaster, during a thousand-bomber raid on Essen on 11 March 1945.

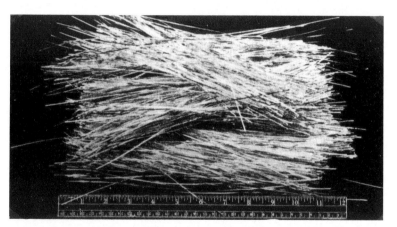

'Window', tin-foil strips ejected in the air to jam and deceive enemy radars.

American Consolidated B-24 ferret-plane equipped for ELINT.

The original 'Enigma' machine used by the Germans to encode messages.

NATO radar installations along a mountain chain near the border of the Soviet Union.

A US Navy McDonnell Douglas F-4J Phantom II and a Soviet Tupolev Tu-95 *Bear*, equipped for ELINT, flying over USS *Nimitz* (CVN-64).

The ELINT variant of the Soviet Tupolev Tu-16 *Badger*. Note the bulges on the nose, under the fuselage and at the tail which cover EW antennae.

The tail of a Soviet Tu-16 *Badger*, photographed from a Royal Navy McDonnell Douglas Phantom FG.1. The arrow points to the RWR antenna radome above the position of the operator. (By courtesy of *Defense Electronics*)

The spy-ship USS *Pueblo*. Note the antennae on her two masts, many of which are enclosed in protective plastic radomes.

The US U-2 spy-plane.

An airbase photographed by a U-2 from 70,000 feet.

Enlargement of the area ringed in the top photograph, demonstrating the U-2's exceptional photographic capabilities.

A battery of Soviet SAM-2 missiles of the type that shot down Powers' U-2.

An obsolete noise jammer (1958) mounted on a truck with a trailer.

Two ships of the Soviet Black Sea Fleet, their masts crowded with antennae.

A typical electronic warfare equipment of the 1960s, a radar warning receiver (RWR). (By courtesy of Elettronica SpA)

A Soviet MiG-25 *Foxbat* of the type which was landed by a defecting pilot in Japan in 1976. (By Courtesy of *Defense Electronics*)

A Soviet spy-ship (AGI) of the 'Primorye'-class. The six ships of this class, which are generally considered to be the most advanced of their type in the world, are able to process and analyse all collected data. Note the many antennae all over her.

A formation of Grumman EA-6B Prowlers. The Prowler carries very sophisticated equipment for tactical electronic warfare. Note the special antenna mounted on the tail.

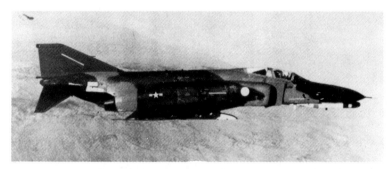

An F-4G Phantom II, 119 of which were converted from F-4E's for use in 'Wild Weasel' missions over Vietnam, fitted with on-board and pod-mounted EW equipment and able to carry AGM-45 Shrike anti-radar missiles (ARM).

A SAM-2 missile from a North Vietnamese battery hitting a US aircraft. (By courtesy of *Defense Electronics*)

managed to crawl its way to its destination without encountering the Home Fleet and, at midday on 13 February, arrived home.

The success of this difficult mission was largely determined by its near-perfect organisation and, above all, by electronic counter-measures.

This episode in naval history highlights an essential factor in electronic warfare: the vulnerability of radar to electronic disturbance. This vulnerability led to electronic countermeasures which are still valid today.

8

Electronic Warfare over Germany

After the serious losses suffered in the Battle of Britain, the German air force was withdrawn from the Western Front and redeployed to air bases in eastern Germany to take part in the Russian campaign, which was designated 'Operation Barbarossa'. The RAF was therefore free to initiate a massive retaliation involving intense air-bombardment of Germany as part of the strategy of destruction which was to secure victory for the Allies.

Long range day bombing had proved unsuccessful, largely due to the vulnerability of bombers to enemy fighters and the inability of RAF fighters to provide more than short-range escort because of limited range, so it was gradually phased out in favour of night raids. Now the roles of the Germans and the British in the 'war of rays' were reversed; this time, the British had to devise fool-proof systems to guide their bombers onto targets and the Germans had to find effective countermeasures.

During the Battle of Britain, the British had noted how difficult it was for the German bombers to hit their targets despite the sophisticated electronic aids at their disposal. The British were faced with exactly the same problems in their air raids over Germany.

How could the British possibly hope to hit their targets in Germany without having accurate radio-electrical navigational and bombing aids? RAF commanders were extremely sceptical about the effectiveness of the first bombings of Germany. Air Marshal Saundby, chief of RAF Bomber Command, commented to his Chief of Staff that, when a squadron of bombers reported that bombs had been dropped on a certain target, one could only be sure that they had been 'exported' towards that target.

Fortunately, the British already had a navigational aids system which had been designed in 1938 but had not been put into production at that time as priority had been given to other projects. This system, called Gee, consisted of three radio transmitters positioned along the coast at 100-mile intervals. They were synchronised to send out a complicated sequence of pulses in a certain order. Bomber navigators had a special receiver which could measure the time-difference between the reception of the pulses coming from the three stations. By referring to a special grid-map of Europe, the navigator could determine his position with a margin of error of about 6 miles, at a distance of 400–500 miles from the transmitters.

Gee was not as easy to fathom out by electronic countermeasures as the early radio-guidance systems employed by the Germans had been. However, it was not long before the Germans noticed that British bombing had become notably more accurate and devoted every effort to finding out what the new guidance system was. By 1942, they had succeeded in doing this and, to counteract it, they built powerful electronic jammers which were called 'Heinrich'. These were installed in ground stations in occupied France, Belgium and Holland and managed to neutralise almost completely the Gee electromagnetic emissions, rendering it practically useless on the European continent.

Various navigational aids systems were tried by the British, following the neutralisation of Gee, but none of them provided the necessary accuracy for bomb-aiming. Finally, they came up with Oboe (Observation Bombing Over Enemy) which was the outcome of careful study of the German 'Knickebein' system. Oboe consisted of a transponder to emit signals, installed on the bomber, and two ground stations a certain distance apart equipped with interrogators (to receive signals). These were called 'Cat' and 'Mouse', respectively. The ground stations were able to measure their distance from the flying aircraft automatically. The Oboe system had considerable success in the Allied bombings of the Krupp works at Essen in December 1942.

After a while, the Oboe system was discovered by the Germans who immediately developed appropriate electronic countermeasures to interfere with its transmissions. So, to replace Oboe, or at least make up for its shortcomings, the British perfected a system called H_2S. This had the dual function of clearly indicating the route and of ensuring greater accuracy in night bombing. Unlike the previous systems, H_2S did not need ground stations: its 'heart' was a recently-developed radar which could be installed on the aircraft. This apparatus utilised a special high-power valve, called the Magnetron, which generated energy of 10,000 watts on a wavelength of 10 cms. For this reason, the new radar was called a centimetre radar to distinguish it from preceeding radars which used considerably longer wavelengths.

The prototype was installed on a test and evaluation aircraft and tested for use in night-fighters. These test flights demonstrated that the new radar was capable of distinguishing built-up areas from countryside and the sea from the land. The test flights were made in

1941 but the system did not enter operational service until much later as the British were afraid that it might fall into German hands and be copied for use on their aircraft. The final decision to use H$_2$S was prompted by the ever increasing losses of RAF bombers in night raids over Germany.

The British high command was also worried about whether the Germans had anti-aircraft radar. Many people were convinced, at least at the beginning of the war, that they did not as no giant antennae like those along the British coasts had been built in Germany or the occupied territories. However, the Germans did, in fact, have anti-aircraft radar right from the beginning of the war but, since they had always been on the offensive, they had not deemed it necessary to build an air defence radar chain requiring huge antennae like those of the British Chain Home.

The increasing number of RAF bombers lost over Germany made it imperative for the British to learn more about German radar anti-aircraft defence in order to devise appropriate countermeasures to neutralise the systems. So, for several months, Allied secret services collected as much information as possible to achieve this end. Frequent reconnaissance flights were made over Germany to search for radar antennae, prisoners were interrogated and all German aircraft brought down in Britain were carefully scrutinised, piece by piece.

In November 1940, an interesting aerial reconnaissance photograph had been taken in the area of Cherbourg in occupied France. It contained an otherwise unidentified object which could be a radar but, since the photograph had been taken from a very high altitude, it was not possible to make a positive identification. It was not until February 1941 that the RAF managed to take a series of photographs from a low enough altitude to be able to distinguish the mysterious object; in fact, it turned out to be the antenna of one of the early German radars called 'Freya' (the Scandinavian goddess of beauty and love) which had first been built in 1939. Its main function was to detect enemy aircraft at the greatest possible range, what we now call early warning.

This radar operated on a 2.5 m wavelength and had a range of about 100–120 miles. Up to a minimum distance of 20 miles, it could detect and track an aircraft, with an accuracy of about half a mile in range and 1 degree in bearing. It was equipped with a transmitting antenna made up of a series of dipoles.

The first 'Freya' radar sets were installed in fixed positions along the northern coasts of France, Belgium and Germany on RAF bomber routes. To compensate for the shortcomings in its secondary AA defence role resulting from its 20-mile minimum range limitation, powerful searchlights were used in association with the radar to illuminate the aircraft. However, this method was too susceptible to poor weather in the area, especially cloud, and so German industry had to produce another radar to produce, more reliably, the information required to direct anti-aircraft artillery and interceptors on to enemy bombers at close range.

The British, having discovered the operating frequency and other characteristics of the 'Freya' radar, were now able to devise appropriate electronic countermeasures to neutralize or, at least, diminish the efficiency of the German radar. Initially, this was fairly easy because all the 'Freya' radars operated on the same frequency (120–130 MHz) which was easily covered by the jammer invented by the British and named 'Mandrel'. This apparatus emitted random noise on the same frequency as that used by the 'Freya', thereby blinding it. 'Mandrel' jammers were installed on special aircraft which accompanied the bomber formations on their raids, helping them to penetrate German airspace. The Germans tried to avoid being jammed by continually changing frequency so the British, to follow suit, had to produce more jammers, of varying types, to cover the different frequencies used.

For a short time, British losses showed a slight decrease but, towards the end of 1942, casualty figures got worse again. The Germans had produced a new, extremely sophisticated radar, called Giant Würzburg, which, operating on a wavelength of about 50 cm (565 MHz), had a range of about 45 miles and was able to measure not only the distance and direction of an enemy aircraft but also its altitude. It also had a very narrow beam and, with all these qualities, was able to provide, with great accuracy, all the essential information for two extremely important functions in air defence: guiding fighters to intercept enemy bombers and directing anti-aircraft gunfire.

Further progress was made in the field of radar when the Germans produced a new apparatus called Liechtenstein BC for installation in night fighters. Although it had a range of only 7.5 miles, it played a very important part in the integrated air defence system. This modular system was made up of numerous stations, each of which had the task of covering a certain zone within a grid covering the west of the Reich. These stations were given the name 'Himmelbett' (four-poster bed.) Each one contained one 'Freya' radar and two 'Würzburg' radars, an operational control room and a communications post. The initial sighting of a British formation was normally made by the 'Freya' which immediately communicated the sighting to the operational control room. A night-fighter equipped with 'Liechtenstein BC' radar would immediately be vectored using one of the 'Würzburg' sets, to intercept the enemy. The other 'Würzburg' tracked enemy aircraft and controlled the laying and firing of anti-aircraft artillery (AAA) once the aircraft were within firing range. All data regarding the positions and altitudes of enemy bombers and intercepting night-fighters were reported on a special table called a 'tactical table' from which the operator could make the necessary calculations for interception. Information such as route, speed and altitude was transmitted, via the appropriate communications post, to the night-fighter pilot who was thus guided to the target from astern wherever possible. When the German fighter was within one or two miles of the enemy plane, the operator switched on his airborne 'Liechtenstein BC' which, having acquired the target, guided the fighter to it. When the fighter was within firing range, the 'Liechtenstein BC' was used to direct the fighter's guns. At this point, the enemy bombers chances of escaping were slim indeed.

This system functioned extremely well and can be considered the forerunner of modern air defence systems despite its limitation of being able to deal with only one bomber at a time. Using this system, a network of air defence was set up along the northern coasts, starting from France and proceeding eastwards. Outside Germany, the systems were positioned at 20-mile intervals whereas in the German hinterland they were spaced at 50-mile intervals.

By the end of 1942, losses of Allied planes to *Luftwaffe* night-fighters and AAA batteries were becoming unacceptable to the Allies. The British stepped up their jamming of the 'Freya' radar sets, frequently sending aircraft equipped with 'Mandrel' jammers along the German coast to prevent the 'Freya' radars from making long-distance sightings. However, when their losses showed no decrease in spite of such measures, it became apparent that the success of the German airdefences depended not so much on the 'Freya' radars as on the pairs of 'Würzburg' radars which the British did not know enough about to be able to jam.

Meanwhile, the Germans decided to try to find a way of protecting the 'Würzburg' radars from eventual jamming by the enemy. They decided to change their frequency continually but this task proved much more laborious than they had expected because they ran up against considerable technical difficulties. However, they managed to devise a system of triple interchangeable frequencies for the 'Würzburg' radars.

While this was being done, the British intelligence service had discovered, near Le Havre in occupied France, the existence of a complex of radar sets, one of which was definitely a 'Freya', while the other two were thought to be the ones their bombers had come up against, the 'Würzburg' sets. Since they knew nothing about the electronic characteristics (frequency, pulse duration etc) of this radar and, therefore, could not devise appropriate electronic counter-measures, they had no alternative but to capture one.

So, on the night of 27/28 February 1943, a company of paratroopers were dropped on the radar station at Bruneval, near Le Havre; their mission was to bring back to Britain the main components of the 'Würzburg' radar sets. Dressed in black, their faces smeared with 'soot', the paratroopers managed to enter the radar station and, after overwhelming the guards, were able to dismantle the 'Würzburg'. The task was soon completed and the company made for the coast a few miles away where a submarine was waiting to take the men and their strange booty back to England. As soon as they had their hands on the components, British technicians set about trying to devise a counter-measure to neutralize the 'Würzburg'.

One night in May 1943, a German Junkers Ju 88R-1, whose crew had decided to defect, landed at a British airfield. This was an unexpected piece of luck for the British who immediately set about examining the JU88's radar. They even went so far as to stage test flight attacks against a British Handley-Page Halifax bomber. In this way, much useful information was obtained, the most important of which was that it had a limited antenna opening of only 25 degrees. Faked combat with the Halifax showed that a slight dive would take the bomber out of range.

The Germans were not resting on their laurels and they too had found ways of neutralising British radar by means of electronic disturbance. They built a jammer for every type of British radar, including their fire-control radars.

The Allies soon came up with a new jamming transmitter, called 'Carpet', which was able finally, to jam the German 'Würzburg' radar sets. It was also installed in the first American Boeing B-17 Bombers, and, thanks to these new electronic warfare systems, Allied bomber losses showed an immediate and progressive decrease: during the bombing of Bremen by the US 8th Air Force, Allied losses decreased by 50 per cent.

However, the worst was yet to come for the *Luftwaffe*. In the late evening of 24 July 1943, the German radar station in Ostend detected a formation of British aircraft approaching from the North Sea. The 'Würzburg' radars in Hamburg also located the enemy formation and communicated to regional headquarters: 'Enemy aircraft approaching at an altitude of 10,000 feet'. That was their last sighting because the echoes on the screens of all the 'Würzburg' sets suddenly grew out of all proportion, totally bewildering the operators who could not believe that there really were thousands of invading aircraft. They eventually reported that their sets were no longer functioning properly and requested instructions.

Meanwhile, the Allied formation had almost reached the outskirts of Hamburg, the anti-aircraft batteries and fighter squadrons having failed to react to the threat due to lack of commands from the 'Würzburg' radars. Partially obscured by something the Germans could not understand, the huge formation, composed of 718 four-engine and seventy-three twin-engine bombers approached the city centre undisturbed. The anti-aircraft defence commanders at Hamburg, frustrated by the lack of data which would enable them to direct their fire and in order not to give the enemy confirmation of the effectiveness of his electronic countermeasures, gave the order to fire blindly at the bombers, but the latter, on reaching their target, successfully carried out one of the most terrible bombing raids in history.

What had happened was that a simple but effective electronic countermeasure had been used for the first time against the 'Würzburg' radars—'Window'. This countermeasure consisted in releasing from the aircraft thin strips of tin-foil of a specific length. To effectively jam enemy radar, the length of the tin-foil strip had to correspond to half the wavelength of the frequency used. Released in bundles which burst open upon ejection, scattering the tin-foil widely, these strips produced return echoes on the ground station radar screens which camouflaged the echoes produced by the aircraft or simulated the presence of huge numbers of aircraft. The radar operators were totally bewildered by the myriad white blips which appeared on their radar screens and were unable to determine the number or position of the approaching enemy aircraft.

The British had come up with this countermeasure a year previously, shortly after the commando raid at Le Havre in which pieces of the 'Würzburg' radar had been captured, but they had hesitated to use it for fear that it would fall into enemy hands and be used against them. Finally, Winston Churchill himself gave orders to use it in the Hamburg raid, planned for July 1943. Orders for the use of this countermeasure by the RAF were given in clear with the codewords 'Open the Window', and so the tin-foil strips were thereafter referred to as 'Windows'; the Americans, on the other hand, referred to them as 'chaff', the term which is now applied to such forms of passive ECM.

This counter-measure had a high degree of success in the raid on Hamburg. Confused by all the false echoes on their radar screens, the German AAA batteries were unable to direct their fire and the fighters no longer received instructions from the ground. Other factors which contributed to the success of the Allies were the excellent meteorologial conditions and the clarity of the images on the screens of their H_2S radar which was due to the sharp contrast between the reflection of the ground and that of the water in the estuary of the River Elbe.

The destruction and casualties caused by the British air raid on Hamburg were enormous. In only two and a half hours, 2300 tons of bombs were dropped on the port and city centre. The intensity of the fires started resulted in a fireball which sucked in huge quantities of air to feed itself upon, draining the city of oxygen, and giving rise to tremendous winds which uprooted trees and swept objects and people into the sea.

Of the 791 bombers used in the raid, only twelve failed to return; this loss-rate was less than a third of the average for the most recent night raids on Germany. Moreover, the chaos which had been wreaked in German air defences had enabled the British to bomb the city with greater accuracy than ever before. The Hamburg raid was undoubtedly the most successful raid ever carried out by RAF bombers and its success must be largely attributed to that simple but effective electronic countermeasure which employed ordinary tin-foil!

It is ironic that the first to have the idea of using tin-foil in this way had been the Germans themselves. They had developed the idea in the course of their research on radar a few years before the war broke out. When Hitler had been informed of the possibility of using tin-foil strips, which the Germans called *Düppel*, he gave the order to break off research and destroy all the relative technical documents. Like the British, he was afraid that the new countermeasure might fall into enemy hands and be copied. Consequently, the local air defence system was taken completely by surprise when the measure was put into effect during the Hamburg raid. On that terrible night, in which

73

tens of thousands of people were killed, nobody had the slightest idea
what was happening not even high-ranking officers of the German air
defence command who, it is reported, gave out the order, 'Don't touch
those strips, they're probably poisonous.'

It was a long time before the German people learnt that those
strange objects raining from the skies constituted the simplest means
of confusing their radar detection and guidance systems. A mere
twenty-five strips were sufficient to create on the radar screen an echo
equivalent to that of an aeroplane; coincidentally, most German radars
operated on frequencies between 550 and 570 Mc/s, the most
vulnerable to jamming, and therefore required a minimum of tin-foil
strips to create interference. During the Hamburg raid, two tons of
these were dropped from each of the aircraft dedicated to this role—a
total of 2000 strips every minute!

Two nights later, a second raid was made on Hamburg, followed by
further raids on other large German cities, all utilizing the new
electronic countermeasure. During the first six of these raids, 4000
individual sorties were flown with a loss of only 124 bombers (3 per
cent of the total), which was much lower than in previous raids. A few
months later, General Wolfgang Martini, head of the *Luftwaffe*
telecommunications service, conceded that the tactical success of the
enemy was absolute.

However, as always happens in electronic warfare, the party was
soon over for the British. After the initial shock, the Germans soon
found ways of getting round the new problem. After a while, the more
experienced radar operators noticed that it was possible to distinguish
between the echoes from the bombers and the 'Window', since the
former moved at a regular speed in a fixed direction while the latter
seemed to be immobile on the radarscopes. The British retaliated by
dropping enormous quantities of tin-foil strips which completely
blanked out the enemy radar screens.

At this point, the Germans decided to produce these precious tin-
foil strips themsevles and, six weeks after the Hamburg raid, put them
into effect, with extremely positive results, in a bombing raid on a
British airbase.

The Germans also came up with a series of electronic counter
counter-measures in an attempt to improve the functioning of their air
defence system. Some of these made use of techniques for distinguishing echoes reflected by aircraft from those reflected by other metallic
surfaces. Another much-used device permitted a radar to change
frequency as soon as it was jammed by the enemy. Yet another system

exploited the Doppler effect: the change in frequency which occurs as a result of the relative movement of the source of a wave and its receiver, thus allowing the radial velocity of a target to be calculated. In this case, the Germans switched from 'video' to 'audio', substituting the radar screen for earphones through which the night fighter pilot could hear a particular sound made by the enemy radar. With this system, changes in the enemy aeroplane's speed were indicated by a change in tone and the operators were able to distinguish even whether the enemy was in a dive or a climb.

These devices, aimed at neutralising or reducing the efficiency of ECM were called electronic counter-countermeasures (ECCM) Nowadays, every military radar has a certain number of ECCMs incorporated into it at the design level; this is usually done by manipulating the circuits of the apparatus or varying its parameters (frequency, pulse rhythms, etc). Many techniques are used today in devising ECCM and, indeed, the possibilities are infinite since, for every countermeasure there is a counter-countermeasure and, for every counter-countermeasure, a counter-counter-countermeasure, and so on.

However, in spite of all the measures taken by the Germans to remedy the situation, night by night they watched their cities being systematically destroyed by RAF Bomber Command. During the summer of 1943, increased use of 'window' by Allied bombers had managed to nullify the German air defence system almost completely during the night and conditions of poor visibility when it relied heavily on the 'Würzburg' radars. So, the best electronics brains in Germany were put to work to find ways of restoring the efficiency of their all-important air defence system.

It was necessary to build a new radar which would use a frequency far removed from those of the 'Würzburg' and 'Liechtenstein BC', whose wavelengths were in adjacent bands in order to avoid the interference produced by the Allied ECM, both active (such as 'Carpet' jammers) and passive (such as 'Window'). Research was conducted at a frantic pace as every night and day that passed could mean the destruction of another German city.

In October 1943, the prototype of the new apparatus was ready and, in the early days of 1944, the new radar, called 'Liechtenstein SN2', was installed on nearly all German night-fighters. It operated on a wavelength of 3.3m, corresponding to a frequency of about 90 MHz, much lower than that of either the 'Liechtenstein BC' or the 'Würzburg', and although the resultant antenna was much bigger and more cumbersome, they had the distinct advantage of being able to cover a 120 degree sector over the nose; such a wide beam was made possible by the higher power of the radar which made it unnecessary to transmit directionally. Now, it would be almost impossible for the British bombers to escape once they had been detected by this radar, but the greatest advantage of the wide beam was that the German fighters would now be able to track down the enemy bombers unaided, once they had received information regarding their formation and approximate route. Detection of the enemy bombers was facilitated by two other factors: the excellent range of the new radar installed on German night-fighters, which was 40 miles, and the fact that the British bombers had recently adopted a new tactic for approaching their target which actually made their discovery much easier for the new German system. Being aware that the German air defence system could only track one aircraft at a time, they had decided to fly bomber streams instead of staggered attacks as they had done previously. However, these huge formations could be detected from the ground even without the aid of radar.

Thanks to the new radar, German defence tactics were completely revised and up-dated since zone defence strictly dependent on ground radar control could be dispensed with. Now, ground control stations had merely to direct the fighters towards a formation and the fighters were then able to operate independently. They penetrated the enemy formation from behind and proceeded to massacre the unfortunate Allied bombers. Previously, once the bombers had got past the defending wall of radars, they only had to contend with AA defences over the target area; now, they were constantly under threat of attack all the way from Belgium and Holland on their way to the target and all the way back to the North Sea after the mission.

The progress made by the Germans in the field of electronics did not stop here. A new RWR was installed on the fighters which were already equipped with 'Liechtenstein SN2' radar. An RWR is an apparatus which has the function of detecting the presence of a radar transmitter; it picks up radar signals but does not itself transmit. The function of these airborne RWRs can be compared to that of the Metox sets installed on German ships and submarines at the beginning of the war. As stated, it has two important advantages over radar: first, it is a completely passive instrument which does not emit electromagnetic energy that could reveal its presence to the enemy and, secondly, it has a greater range than a radar since it receives emissions from the enemy radar before the latter is able to receive a signal returned from the platform on which the RWR is installed. In practice, this meant that the RWR installed on board German fighters were able to receive the radar emissions of the Allied bombers at almost double the distance at which the bombers' radars were able to detect the German fighters. Consequently, the fighters had plenty of time to plan their manoeuvres. The RWRs were also able to guide the fighters to the enemy formation as, although unable to measure their distance from the enemy radar, they gave a fairly accurate indication of the direction from which the transmission was coming. Moreover, being completely passive, the RWR was immune to disturbance from the tin-foil strips which had caused so much trouble on other occasions!

By the beginning of 1944, the Germans had two types of RWR installed on their fighter planes. One, the 'Naxos' was able to pick up British H_2S radar transmissions. Since H_2S radar had been, for the moment, only installed in aircraft of the specialised RAF Pathfinder Force (PFF) which had the task of marking the targets to be bombed by dropping phosphorus flares for target illumination, the 'Naxos' guided the German fighters directly to these aircraft, which had such a vital function in British strategy.

The second German RWR the 'Flensburg', was tuned to receive transmissions from another type of British airborne radar 'Monica', which, installed in the tail of RAF bombers, gave warning of the approach of enemy fighters to enable the bombers to take the appropriate evasive action. The Germans had found one of these radars among the wreckage of a shot-down enemy bomber and had the bright idea of exploiting its transmissions to get the British right by the tail, as it were!

The 'Flensburg' RWR constituted an authentic self-guidance system leading the fighter right on to the enemy's tail, where their radar was installed. The 'Flensburg' RWR consisted of a comparison receiver and two identical antenna installed in the front of the fighter at an angle of 60 degrees away from each other. When the antenna on the left received a signal showing up on the radarscope, it simply meant that the bomber was to the left of the fighter whereas, if the antenna on the right received a signal, it meant that the bomber was to the right. When the two antenna intercepted a signal of equal intensity, it meant that the enemy bomber was dead ahead. With this exceptional radio electrical device, the *Luftwaffe* obtained, initially, outstanding results.

In 1944, the total destruction of Berlin was prevented largely due to this progress made by the Germans in the field of electronics. The efficiency of the German night-fighters, with the support of the well-organised anti-aircraft artillery, prevented RAF raids from causing destruction on the same vast scale as that of Hamburg.

During this period, RAF losses rose considerably and morale dropped proportionately. Many of the best British pilots had reached the limits of their endurance and often, at the slightest sign of danger or difficulty, dropped their bombs in the sea or open country. As soon as they heard the noise of the inexorable approach of enemy fighters, the bombers' terrified gunners began to shoot at anything they saw or imagined they saw and sometimes they shot down one of their own aircraft by mistake.

This chaotic state of affairs reached its culmination on the night between 30/31 March 1944 when German fighters, guided by their RWRs, zeroed in on an RAF bomber formation over Brussels and engaged it in an air battle which went on all the way to Nuremburg, the target of the raid, and all the way back. The Allies lost ninety-five of the 795 bombers sent out on that mission, while another seventy-one returned to base badly damaged and twelve more crashed on landing. The final toll was 115 bombers and 800 highly-trained crew members lost. It was a great victory for the Germans; one pilot claimed seven kills and many had two or three. The victory can be largely attributed to German supremacy in the field of electronic warfare at that stage of the war.

The situation was becoming extremely critical for the RAF until, by an unexpected stroke of luck, it was able to remedy the situation by appropriate electronic retaliation. At dawn on 13 July 1944, one of the most modern German night-fighters, a Junkers Ju 88G-1, landed in England as the result of a navigational error. It was equipped with all the latest electronic equipment (SN2 radar, 'Flensburg' RWR and some highly efficient new radio sets), except the 'Naxos' which, fortunately for the Germans, had not yet been installed on that particular aircraft. British experts immediately started a thorough examination of all the equipment and were utterly dismayed when they realised what the purpose of the 'Flensburg' was. Instead of protecting them from enemy fighters, the tail-mounted radar was attracting them like flies to meat and was enormously facilitating their attack.

To convince the incredulous RAF commanders, a trial flight was organised in which seventy-one Lancaster bombers, all equipped with tail radar sets, were ordered to fly towards Germany as though on a real mission. A Ju 88, piloted by a British crew, took off and all the bombers were then ordered to switch on their electronic equipment. The 'Flensburg' RWR managed to pick up the electronic emissions of the British radar at a distance of nearly 50 miles and, without turning on its own radar, the Ju 88 was able to come up behind the Lancaster bombers and get into the very best position for firing at them. There was no doubt about the efficiency of the 'Flensburg' radar and all radar equipment was promptly removed from the tails of RAF bombers.

Meanwhile, huge quantities of tin-foil strips, cut to the correct size for the wavelength of the 'Liechtenstein SN2', were produced and, towards the end of July 1944, this new 'Window', was already in use. British losses in night raids over Germany began to show a significant decrease as a result both of the use of the new 'Window' and the removal of the tail radar from their aircraft.

The Germans then tried other technical measures to reduce the electronic disturbance caused by the 'Window', such as modifying their radar antennae. When the British became aware of this, they started using very long metal strips (up to 400 feet) attached to little parachutes, each capable of simulating the echo of a large aeroplane. The Germans were obliged to modify their radar further in an attempt to eliminate the effect of the new British countermeasure.

Meanwhile, as the war dragged on, the Germans were experiencing various problems, such as the increasing losses of their courageous and highly skilled pilots, the difficulty of training new ones to replace those lost and the increasing scarcity of fuel.

At the same time, the British were becoming more and more convinced that every effort must be made to neutralize the electronic components of German air defence. To this end, they set up special squadrons, mainly composed of Short Stirling aircraft, equipped with 'Mandrel' jammers capable of jamming the 'Freya' early warning system. The Stirlings also carried huge quantities of 'Window' which enabled them, singly or in pairs, to cause echoes to show up on the enemy radar which falsely indicated the presence of large formations of bombers. This would distract the attention of German air defence from the real bombers which were attacking elsewhere.

However, before the war came to an end, German industry managed to come up with two new radars against which these Allied ECMs were ineffective. The first was called 'Neptun' and worked on a combination of six frequencies from 158 to 187 MHz, corresponding to wavelengths between 1.9 and 1.6 m, which could not be jammed by 'Window'. The second was a radar called 'Berlin'. It was a revolutionary invention, in its time, working on a centimetre wavelength. Its antenna was no longer a complex system of dipoles installed on the outside of the aircraft but a parabolic antenna inside the nose. Only a few models of the 'Berlin' radar were manufactured before the end of the war.

The Junker 88G-7b was equipped with the 'Neptun' radar as well as with a device capable of distinguishing enemy from friendly aircraft; this was the forerunner of IFF (Identification Friend or Foe), which is installed on all modern military aircraft, and can distinguish enemy from friendly aircraft. It was also equipped with a radio-altimeter, a radio-compass, a secure navigation receiver that printed out in clear Morse code the aircraft's position as transmitted by a ground-station, blind landing instrumentation and two new HF and VHF radio sets. Since the 'Neptun' relied on beamed high-powered transmissions and the teleprinter signals had good 'break through' qualities, like Morse code, the systems were highly jam-resistant. The Junker 88 G-7b also carried the 'Naxos' while the 'Flensburg' RWR was replaced with an infrared (IR) ray device ('Kiel') that reacted to heat radiations from 'hot spots' such as the exhaust of the enemy aircraft's engines.

During the last months of the war, both sides used the trick of creating false targets. Radar is not capable of determining the form or nature of the object detected and so it was easy to use various metal surfaces to create an echo which, in the right circumstances, would be taken for that of an aircraft, ship, etc.

False targets were used extensively by the Germans in the Berlin area to prevent the total destruction of their capital city. They set up numerous metal targets in the nearby lakes, hoping to deceive the Allied bombers, which used the H_2S radar for blind bombing.

These and other more sophisticated devices were used by both contenders in the final stages of the war. In the skies over Germany a continuous struggle was going on between radar, electronic counter-measures and their counter-countermeasures. It was certainly one of the most dramatic challenges in the whole of World War Two both on a scientific level, the opponents being equally matched in technical expertise, and on an operative level, both sides fighting with desperate determination and great skill and courage.

After the United States entered World War Two, the number of aircraft participating in each battle grew considerably. During the final months of the war, Germany was being bombed daily by forces of no less than 1000 bombers, escorted by between 600 and 700 fighters, and nightly by almost the same number of RAF bombers.

The struggle between the fighters themselves, tactics for day or night fighting, the organisation and efficiency of air defence, the continuous improvements made in detection, guidance and ground control were all very important factors which rendered the outcome of the struggle uncertain right until the final day. Allied air losses over Germany were extremely high; it is thought that between about twelve and fifteen thousand aircraft were lost.

As in the Battle of Britain, the struggle between radar and electronic countermeasures played an extremely important role in the air battles over Germany, first favouring one side, then the other according to the efficiency of the new electronic devices introduced and the surprise element which would catch the enemy off-guard.

9

Electronic Deception in Operation 'Overlord'

With the invasion of Normandy, code-named 'Operation Overlord', electronic countermeasures played for the first time in history an integral part in strategic plans. They were, in fact, one of the more important elements in the overall plans drawn up by the Allies for one of the most complex military operations in history.

This invasion was of vital importance but its success was by no means certain. It is well known that the critical stage of a landing operation is the period during which the troops are being carried from the ships to the beach by the landing craft. This period can last for several hours and, if the enemy is in a position to attack them on landing, the result can be a massacre as the troops are extremely vulnerable when they first 'hit the beach'.

It was, therefore, of the utmost importance for the Allied Command to deceive the Germans about the actual landing area, and thus delay the movement of their strategic reserves towards the area to counter a possible landing. It was decided to try to convince the Germans that the landings would take place near Calais when, in fact, they would take place on the beaches of Normandy. The electronic plan was extremely complex and, of course, top secret. It would be put into effect several days before D-Day and involved a combination of actions, some real and some fake.

The beaches of Normandy which the Allies chose for their landings were heavily fortified, as was the entire northern coast of Europe. Field-Marshal von Runstedt was in command of the sixty divisions which manned the so-called Atlantic Wall, the system of fortifications which ran from Holland to the Bay of Biscay. The already famous Field-Marshal Rommel was in command of the sector between Holland and the Loire.

The Germans, of course, knew that the Allies were planning an invasion of Europe and that they would certainly land somewhere in northern France. Von Rundstedt was convinced that they would land at Calais; Rommel, on the other hand, surmised that the landing would take place on the beaches of Normandy.

German government leaders were also divided in their opinions as to where the landing would take place. These differences of opinion were due to a series of deliberate actions taken by the Allies to try to deceive the Germans into thinking that they would land at Calais.

The Germans, naturally, did all they could to complicate the Allies' plan, conducting a propaganda campaign featuring the impregnability of the Atlantic Wall. In a radio transmission in March 1944, about two months before D-Day, they claimed that their radar chain surrounding the whole of Germany was so efficient that each and every enemy craft would be under constant observation and, with these assets at its disposal, German defence could operate with extreme speed and efficiency.

The Allies were well aware of the fact that the Germans had installed at least 120 radars along the northern coast of France for the purposes of detecting British convoys in the Channel and directing coastal gunfire. Through photographic and electronic reconnaissance, they knew all about the entire German radar chain, which consisted of radars positioned at 10-mile intervals and, in some parts of the coast, every half a mile.

British electronic experts had begun to devise detailed counter-measures well in advance. They had chosen a stretch of beach along the Scottish coast which closely resembled the coast of Normandy and had installed three captured German radar sets, representing the three main types of radar guarding the beaches of Normandy. Every day, aircraft, naval vessels and landing craft furnished with EW equipment carried out practice landing operations on the Scottish beach. Officers who were experts in electronic warfare umpired the manoeuvres to determine how successful the 'invaders' had been in jamming the enemy radar. From these exercises, there evolved a detailed table of equipment requirements for the ships and aircraft that were to take part in the invasion. Every skipper and pilot was given explicit instructions regarding what he had to do on D-Day.

Basically, the plan involved two main actions. The first was to jam German radar in the Normandy area to prevent detection of the approaching naval force. The second action involved deceiving German radar in the Calais area by simulating the presence of a large fleet sailing towards Calais. Other supportive measures were planned to operate in conjunction with these two actions. Intense fictitious radio traffic was instigated in the Dover area to give the impression that the trooops were assembled in that region, ready to invade in the Calais area. Rumours and false reports were spread by undercover agents to further confuse the issue. Troops were concentrated in irrelevant areas and, finally, enemy radio communications were routinely jammed.

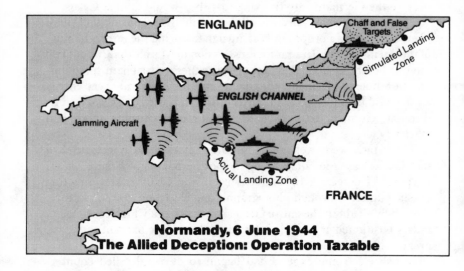

The invasion was scheduled for 06.30 hours on 6 June. On the night of 5/6 June, a huge fleet, composed of about 2700 ships of all types with hundreds of thousands of men on board, weighed anchor from various ports in south-west England and sailed slowly towards the coast of Normandy. At the same time, twenty aircraft, equipped with powerful 'Mandrel' electronic jammers, flew along the south coast of England, at an altitude of about 18,000 feet in order to mask the presence of the approaching ships from German radar installed on the Normandy coasts.

Almost simultaneoulsy, flotillas of small vessels left various harbours in the vicinity of Dover carrying special metallic plates and towing buoys and metallized balloons to create radar echoes of equal strength to those produced by large warships. Shortly afterwards, several aircraft, flying overhead, dropped huge quantities of 'window', or 'chaff', to give the impression of a convoy of ships approaching the French coast in the vicinity of Calais.

As the hour for the landing drew near, all the shipborne electronic warfare equipment was switched on simultaneously, producing sufficient interference to neutralize the efficacy of German coastal gun control radar.

All went according to plan and the invasion of Normandy was a great success for the Allies. The effectiveness of the electronic warfare planning ensured that the German strategic reserves were kept out of the way until the Allied troops had safely established their beach-head.

Allied losses were contained as there was no major confrontation with the enemy during the actual landings. The confusion created by the electronic measures continued the day after the landing, inducing German leaders, including Hitler, to make serious errors of judgement and take wrong decisions.

The success of the electronic countermeasures employed in the invasion of Normandy is best expressed in the words of Winston Churchill:

Our deceptive measures before and after D-Day, were planned to provoke confusion of ideas, their success was admirable and the consequences long withstood during the battle.

10

Electronic Warfare in the Mediterranean

Generally speaking, electronic warfare in the Mediterranean never reached the levels of the Battle of Britain or the Atlantic conflicts. Nevertheless, radar played a crucial role in some phases of the conflict, such as the struggle between British and Italian naval forces, following Italy's entry into the war on 10 June 1940.

The main task of the British fleet in the Mediterranean, which was mostly stationed at Alexandria in Egypt, was to keep open sea-lanes to Malta, an important air and naval base. The Italian fleet, on the other hand, had to keep open sea-lanes to Libya, Albania and the Dodecanese to supply the troops operating there with the necessary provisions.

Naturally, both fleets also had the task of preventing the enemy from achieving its objectives. To this end, a long series of maritime operations was carried out night and day, each side trying to protect its own convoys and attack those of the enemy.

As we have seen in previous chapters, radar had been secretly researched and developed in many countries, including Great Britain and Italy, before the outbreak of World War Two. However, whereas in Italy radar had remained at the prototype stage (for reasons explained previously), in Great Britain it had already gone into production and service, largely due to the pressing requirement for air defence.

At the beginning of the war in September 1939, there had been a great technological and industrial mobilisation in Great Britain which led to the installation of the first radar sets. These had proved extremely useful during the Battle of Britain for sighting German aircraft at night or in conditions of poor visibility at a much greater distance than that permitted by the naked eye.

Nevertheless, during the first encounters between the two fleets during the summer of 1940 (the battle of Punta Stilo on 9 July 1940 and the battle of Cape Spada on 19 July 1940), the Italians had had the distinct impression that the British did not have radar equipment on board their ships. This observation led to no measures being taken by the Italian authorities to speed up tests being made on radar prototypes at the Naval Academy Research Institute of Livorno. Instead, they decided to utilise the few electronic experts they had at their disposal to deal with problems which they considered to be more urgent.

In January 1941, an extremely efficient German air corps, the X CAT (X Air Corps) entered the war in the Mediterranean to give air support to the Axis forces there. The X CAT arrived in Sicily in the middle of an important British air and sea operation, Operation Excess, which involved transporting supplies to Malta. On the eve of important Italian naval missions, German radio operators would board the Italian flagship to ensure radio contact with the X CAT. As we shall see, this had undesirable consequences for the Italian Navy.

The British convoy, coming from Gibraltar, was escorted by sixteen warships, at least five of which had radar on board. They were to be joined in the Straits of Messina by a naval formation coming from Alexandria, Egypt, composed of battleships, cruisers and destroyers, and by two cruisers coming from the Aegean, HMS *Southampton* and HMS *Gloucester*, neither of which had radar on board.

The British convoy and its escort units were continually attacked by Italian air and naval forces and by the German X Corps aircraft. Of the British ships, *Southampton* and *Gloucester* got the worst of it. Not having radar on board, they were unexpectedly bombed by a dozen aircraft which they had failed to detect during their approach. *Southampton* was so badly damaged that it had to be scuttled whereas *Gloucester*, although it too was badly hit, managed to reach Malta.

After these events, the need for radar was seen to be of the utmost importance as a means of detecting approaching enemy aircraft. More and more British ships were furnished with radar equipment and the Italian Navy soon began to feel the effects. During the bombardment of Genoa by a British naval unit based at Gibralter on 9 February 1941, the British ships were able to detect Italian reconnaissance aircraft before they were themselves sighted. The ships therefore had time to change course while fighters were dispatched to attack the Italian aircraft.

During the battle of Cape Matapan, which took place on the night of 28 March 1941, the British were able to intercept, via radar, an Italian formation made up of three 10 000-ton cruisers and four destroyers at a distance greater than the maximum permitted by nocturnal visibility.

On the evening of 27 March, Admiral Cunningham, Commander-in-Chief of the Mediterranean Fleet, having been informed of an imminent attack by Italian naval forces against enemy merchant convoys heading for Greece, gave orders to send out a naval unit composed of one aircraft carrier, three battleships, four cruisers and nine destroyers. Only three of these ships – the aircraft carrier *Formidable*, the battleship *Valiant* and the cruiser *Ajax* – had radar equipment on board. It was radar type 279, a model designed mainly for long-range air search and for guiding fighters onto air target.

The Italian naval formation left port on the evening of 26 March, with German radio operators on board to make the necessary radio contacts with the German X Air Corps, as requested by the Italian command. Both the British and Italian naval formations headed for the island of Crete, arriving on 28 March.

The numerous actions carried out that day are beyond the scope of this book which will limit itself to describing facts and events related to the nocturnal encounter of Cape Matapan, in which radar played the leading role.

At 20.15 on 28 March, the radar operator on the British cruiser *Ajax*, part of a naval formation also comprising the battleships *Warspite*, *Valiant* and *Barham* and the aircraft-carrier *Formidable*, noticed something on the radarscope at a range of approximately 6 miles to port of the ship's bow. Admiral Pridham-Wippell, commander of the cruiser formation, merely reported the radar contact to his Commander-in-Chief and continued on the same course. Admiral Cunningham, on board *Warspite*, decided to sail with the main formation to investigate this radar contact.

About an hour later, *Valiant* established radar contact with an unknown ship which seemed to be stationary at a distance of 6 miles. She maintained radar contact until it was at a distance of about 8000 yards and could see the stationary ship. At the same time, *Warspite*, the flagship, sailing to port of *Valiant*, suddenly sighted two large enemy cruisers which were almost dead ahead of the British battleships. *Valiant's* radar equipment was immediately directed at these new targets showing their range to be 8500 metres, and the three battleships were given coordinates to aim their guns. Steadily the British battleships closed up on the targets until they were at a range of 2800 yards. At this point, the light destroyer *Greyhound*, which was escorting the battleships, trained its powerful searchlights onto the two ships which turned out to be the *Zara* and the *Fiume*: These were en route to assist the stationary ship *Pola* which had been hit by an air-launched torpedo. The British battleships, opened fire at point blank range with all their heavy guns. In two or three minutes, *Zara* and *Fiume* and the destroyers *Alfieri* and *Carducci* were reduced to smoking metal. However, in spite of the onslaught, the *Alfieri* somehow managed to approach the enemy and open fire from behind a smoke screen, causing the British battleships to withdraw. The four Italian ships which had been hit sank during the night, while the *Pola*, which had been lost sight of by the main formation, was sunk at dawn by torpedoes launched by the British destroyers *Jervis* and *Nubian*.

Admiral Cattaneo, Italian Fleet commander, the various ships captains and most of the crews perished that night.

Other tragic episodes resulting from shortcomings of Italian industry followed the battle of Cape Matapan. Although orders had been given for the immediate construction of fifty 'Gufo' radar sets to be installed on board Italian ships, Italian industry lacked the know-how to produce vital electronic components such as the cathode-ray tubes which, in the case of the prototypes, had been supplied by the Americans. Production of these radar sets was, therefore, delayed.

For several months, the British, who had by now acquired considerable experience in radar-controlled night-firing, were practically unhindered in their attacks on Italian convoys heading for the ports of North Africa.

One of these nighttime tragedies was when Italian torpedo-boats made a courageous attempt to force their way through the harbour defences at the port of Valletta in Malta on 26 July 1941. A small radar installed on the island picked up the incoming Italian ships at a distance of approximately 20 miles. The ships sailed right into the alerted coastal defences, which pinpointed them without difficulty and opened fire.

While the Italian navy was waiting for the 'Gufo' radar sets to be built, Germany loaned them their Metox radar receivers which had proved so valuable in the Atlantic. These were installed on board the larger Italian ships, on destroyers used to escort convoys and on submarines. Other radar units, such as the De Te, model Fu. Mo. 24/40G, which had a range of about 20–25 kms, were also installed on various units. The Metox interceptor gave immediate proof of its usefulness. Thanks to their ability to detect the interceptor's electromagnetic transmissions coming from enemy radar units, the Italians were able to make the necessary preparations to meet surprise attacks.

If the Italians had given impetus to the development of electronic warfare systems, operations in the Mediterranean would, no doubt, have had quite different results. The widespread installation of RWR, which, as explained in previous chapters, have a greater range than radar, would have enabled them to detect the enemy prior to being discovered and to jam the enemy radar in the process. The holocaust of Cape Matapan, in which 2300 sailors perished in the Aegean Sea, was the price the Italians had to pay for their lack of electronic warfare equipment which would have enabled them to avoid that deadly encounter.

When the first 'Gufo' radar sets were finally installed on board Italian ships, it was too late; Italy was already losing the war. The lesson to be learned from these events is that, in the field of radar and other electronic warfare equipment, prototypes will remain at this stage unless research is backed up by an adequate electronic industry possessing advanced technology and the ability to build components for mass production. (Today, however, Italian industry is among the most advanced in the world in the field of electronic warfare.)

Radar and electronic warfare played a fairly modest role in air operations carried out in the Mediterranean area during the first years of the war. The Italian air force (*Regia Aeronautica*) unlike the *Luftwaffe* in the Battle of Britain, had felt no pressing need for radio-navigation equipment to guide its bombers. Moreover, the island of Malta, where only a few old-fashioned British aircraft were based, was not considered to be a primary target.

However, the situation changed in 1942 when the island of Malta became an obstacle to the shipping lanes of the Axis, who had to take supplies to ports in North Africa. The lack of radar and electronic equipment was heavily felt during the Axis bombings of the island.

Elimination of the British early warning radar on the island of Malta was given top priority by the Italian high command, particularly in view of the planned invasion of the island. The equipment researched and developed at the Air Force Institute of Radio-Technology at Guidonia, near Rome, in collaboration with the Army and Navy and under the direction of Professor Latmiral, was both original and ingenious, combining a receiver and a jammer. It employed the then widely-used super-reactive technique, no longer in use today due to limits of precision controllability and its tendency to generate very heavy interference which causes problems for other nearby friendly units. It was precisely this negative characteristic that was exploited by Professor Latmiral in order to generate noise waves at the same frequency as the victim radar but with a different pulse repetition rate so as to jam reception. The first sets operated on a frequency of 170–220 MHz radiating at a power of 10–20 watts.

Professor Latmiral's receiver-jammer was used for warfare purposes only for a short period as Italy signed the armistice on 8

September 1943. Therefore, it had little influence on the outcome of the war in the Mediterranean where Italian aircraft had to operate without radar or equipment to protect them from the British radars. To help their Italan allies, the Germans sent troops and tanks to the North African front. At the same time, they began to install radar equipment on the Italian mainland and islands, keeping this a secret from the Italians themselves; the number of radar sets installed in Italy, mainly 'Freya' and 'Würzburg' models, was gradually increased.

When the Allies decided to invade Sicily, it was agreed that they must first find out as much as possible about German radar equipment installed in the south of Italy. RAF Wellington bombers equipped with radar receivers were sent on electronic reconnaissance missions. Flying at a low altitude, they tried to pick out the stronger radar signals and, when they intercepted one, flew towards its source to see if its intensity increased. In this way, they were able to establish the position, or at least the direction, of the radar station. These missions were extremely risky as the Wellingtons often had to fly very close to the enemy radar stations and were often exposed to deadly barrages from defending AA gun batteries. Nevertheless, the Wellingtons managed to locate nearly all the German radar stations installed in Sicily and southern Italy, even though they did not have direction-finders on board.

Towards the end of 1942, more modern and better-equipped aircraft were sent from the United States to replace the slow and antiquated Wellingtons. These were four-engine B-24 Liberator bombers. They carried a civilian specialised in ELINT operations to operate the receiving equipment on board for such missions. To avoid being spotted from the ground during their nocturnal missions, the aircraft were painted black and their engine exhaust flames were shrouded. They had a RWR antenna installed on the tail, pre-tuned to the frequency used by the German night-fighters. When the radar emissions from a German fighter were intercepted, a red warning-light on the flight panel would come on. To determine the direction of the ground-based radar signals, two radar receiver antenna were installed one on each of the wing tips of the B-24s; these were tuned to the frequencies of the 'Freya' and 'Würzburg' radars. When a radar signal was intercepted, the operator could hear the enemy pulse repetition frequency in his earphones. He would then instruct the pilot to manoeuvre the B-24 until the pulses from the two receivers were of equal strength. This meant that the B-24 was now heading directly for the transmitting station and the operator could thus determine its bearing with a fair degree of accuracy. However, this method was not without its shortcomings either. It did not afford sufficient accuracy in establishing the exact positions of the radar stations and also meant that the aircraft had to fly dangerously close to them with consequent exposure to their protective AA gun batteries.

These problems were solved when the first direction-finders (DF) were installed on board the B-24s. These instruments were capable of determining the direction of arrival of electromagnetic emissions from a radar station. With the DF on board, the B-24s no longer had to fly towards the German radar stations in order to determine their bearings; it was sufficient to fly parallel to the coast, measuring the bearing of a radar station at set intervals. The position of the station was then determined by triangulation.

Using this method, the Allies succeeded in locating not only all the radar stations which the Germans had installed in Italy, but also those installed along the south coast of France. The value of this detection was appreciated when the Allies landed in Sicily, at Salerno, and in Provence. Before each operation, they bombed coastal radar stations which would otherwise have warned the Germans of their approach.

However, in September 1943 in spite of all these precautions, the Allies were confronted with a new, unforeseen problem of electronic warfare . The Germans had devised a new, deadly anti-ship weapon in the form of a large rocket-powered bomb which could be radio-guided to its target by the aircraft which had launched it. The bomb, the Henschel Hs 293, was guided by a four-tone radio-control system on board Junkers Ju 88 or Heinkel He 177 bombers; the operator on board the aircraft could control the trajectory of the bomb by sending radio signals to command the movable winglets installed on the bomb's fuselage.

Proof of the precision and destructive power of this new, revolutionary bomb was furnished by an incident which took place in the Mediterranean. On 9 September 1943, the day after Italy had signed the armistice with the Allies, an Italian naval squadron made up of three 35,000-ton battleships, *Roma*, *Vittorio Veneto* and *Italia*, six cruisers and two squadrons of destroyers, was steaming towards the port of La Maddalena in Sardinia. As soon as they had sighted the island of Asinara, the ships steered 45 degrees to port to get on course for entry into the estuary of La Maddalena. But, just as they were passing the straits of Bonifacio, news arrived from the naval high command in Rome, the Supermarina, that the port of La Maddalena had been occupied by the Germans. They were therefore ordered to change course and head for Bona in Tunisia.

While the Italian ships were sailing back out to sea, the Germans, who, no doubt, had intercepted and deciphered the Italian's radio communications, immediately sent out aircraft to bomb them. It was evident to those on board the ships that the bombs launched from the Junkers bombers were rocket-powered and radio-controlled, as they dropped at a greater speed than conventional bombs and fell directly on target. Two bombs hit the battleship *Roma*, Flagship of the squadron, causing such serious damage that it very quickly broke in two and went down with nearly all the crew.

A few days later, on 14 September 1943, the deadly new bombs made another appearance during the Allied landing at Salerno when they devastated the British battleship *Warspite* and the American cruiser *Savannah*. The crews of both ships reported that the bombs launched from the German bombers seemed to be radio-controlled.

The British and American naval high commands, extremely worried by the accuracy of these new bombs, urgently summoned technical experts from their respective countries to try to devise a way of countering this new threat.

The first idea they came up with was to install special receivers on board Allied ships operating in the Mediterranean; these would pick up, record and analyse the electromagnetic signals coming from the aircraft which launched and then guided the bombs. This would have to be done while the ships were under attack and the bombs were being guided towards them—no easy matter!

Some time later, the Allies had the good fortune to get hold of an Hs-293 which had fallen in the coastal waters of Libya without exploding. A thorough examination of this bomb and analysis of intercepted emissions showed that two of the four tones used to guide the missile were above the range of audible frequencies and even above the frequencies which could be recorded by the equipment then in use. While new recorders capable of detecting these high frequencies were being hurriedly produced, knowledge of the other two frequencies employed was sufficient to improvise electronic equipment capable of jamming the enemy's system. The first examples of this new equipment were installed on board the American destroyers *Davis* and *Jones* and proved successful in jamming the radio-guidance of the German bombs.

Technical experts later discovered that the two inaudible frequencies were simply ultrasounds and that the four tones were merely commands to move 'higher', 'lower', 'more to the right' and 'more to the left'. After this discovery, the Allies could intervene with false commands to lead the deadly rocket-bombs off course. As soon as the Allies had produced a jammer capable of doing this, no more of their ships were hit by radio-controlled bombs and these were no longer used by the enemy, having already outlived their usefulness!

11

The Pacific Theatre

Electronic countermeasures played a less important role, and were somewhat different in character, in the Pacific than in the Northwest European and Mediterranean theatres of war. This can be attributed mainly to the low level of Japanese technology and to the geographical characteristics of the area.

Japanese radar equipment was decidedly inferior to that of the Germans and the Allies, both in quantity and quality, and never posed a real problem for US forces. However, the vastness of the Pacific required a large number of suitable devices in order to carry out electronic espionage and, thereby, find out how many radar sets had been installed by the Japanese and what type of equipment was employed. This was a difficult task as many radar stations were situated at great distances from American bases.

The first episode of American electronic espionage in the Pacific took place, in March 1943, in the Aleutians, a chain of rocky islands running from Alaska to the Sea of Japan, some of which had fallen into Japanese hands. Since the Pearl Harbor disaster on 7 December 1941, the Americans had been conducting systematic photographic reconnaissance missions over the Japanese-occupied islands in order to prevent further such surprise attacks. During these missions, a photograph had been taken of the island of Kiska which showed that the Japanese had recently erected two structures, which looked like huge billboards, on the top of the highest mountain. Examination of this photograph by electronic warfare experts showed that they were, in fact, radar antennae for long range air search.

Further electronic reconnaissance flights, in which special receivers were used, collected data regarding frequency, pulse width and other parameters, on the basis of which it was possible to establish not only the type of radar, but also its coverage and the emission diagram of the antennae.

This information proved extremely valuable for the Americans when they began to bomb the island because analysis of the radar had shown that there was a 'blind' sector where the radar beam was in the 'shadow' of one of the mountain peaks. Consequently, the American pilots could approach the island without being detected by the radar installed there.

This episode constitutes an important chapter in the history of electronic warfare since it showed how valuable that type of exploratory mission could be for military operations. The aircraft used in these missions were called 'ferret' planes as, like the ferret, they keenly searched out their prey, which, in this case, was radar.

This type of mission was not limited to aircraft, however. Many warships were equipped with suitable instruments and sent out on similar missions in the Pacific Ocean. The range at which these ships could pick up enemy radiations was decidedly inferior to that of aircraft which have the advantage of altitude. On the other hand, the ships could stay longer in the area under observation, which gave electronic specialists on board more time to pick up, record and analyse the radar emissions.

For such tasks, the US Navy also equipped many large aircraft with intercept receivers and DF. The best-equipped aircraft for such 'electronic reconnaissance' missions was the four-engine Consolidated-Vultee PB4Y2 Privateer, a maritime version of the company's famous B-24 Liberator bomber. Each Privateer carried a dozen operators on board as well as the crew and could be considered a real radar interception centre. The aircraft were easily recognised because the fuselage was covered with radomes made of a special synthetic material, which covered the numerous antennae installed to pick up enemy radar signals. Because of their ugly appearance, individual aircraft were given names of the strangest and most horrible animals.

The Privateer performed an invaluable service throughout the whole Pacific war. Two of them are particularly worthy of mention. They patrolled the whole of the south Pacific, from Australia to the island of Borneo, ferreting out radars, which were subsequently bombed, and supporting naval forces against Japanese merchant traffic.

Submarines were also equipped for this type of mission. They provided an ideal platform for transporting equipment used in electronic espionage and, being able to lie in wait for long periods of time with only their conning-towers above water, they managed to listen to and record enemy radar transmissions and communications. These interceptions were then used to prepare appropriate electronic countermeasures, and often even enabled the submarines to avoid surprise attacks by the enemy.

One such case of 'early warning' was when an American submarine, which had been seriously damaged in combat, was being escorted to base by two others. While the small formation was sailing through the mist, one of the escort submarines intercepted radar emissions from a Japanese aircraft which was flying nearby. Given their precarious situation, the Americans were caught on the horns of a dramatic dilemma: whether to submerge or remain on the surface. If they submerged, they would lose the crippled submarine but, if they remained on the surface, all three might be sunk.

The commander of the EW-equipped submarine decided to use the receivers on board to explore the whole range of frequencies employed by American airborne radar in the hope of finding one of their own aircraft in the vicinity. The exploration was successful and the radar operator on board the submarine was able to give the pilot of the friendly aircraft information which would enable him to find and attack the enemy. While both aircraft were flying towards the unfortunate submarines, the Japanese pilot noticed that he was being followed and dropped his bombs prematurely, completely missing the target. The American pilot was then able to shoot down the enemy aircraft right before the eyes of the bewildered submarine crews!

When the war in the Pacific reached a turning-point in favour of the Americans, who were able to launch strategic and tactical air raids and sea-landings against Japanese-held territory, electronic warfare took an active, and rather different, part in the various operations. For example, during their invasions of the well-fortified Japanese-held islands, the American bombers were generally equipped with systems for electronically neutralising the Japanese radars, such as jammers or chaff, as they had done in their bombing of Germany. Later, each American wing was equipped with converted bombers which carried extra fuel and jamming equipment instead of bombs. Because of their spine-like antennae, these aircraft were nick-named 'porcupines'. Flying over the target with the first wave of bombers, they jammed or neutralised Japanese anti-aircraft gun control radar. They remained in the area until the last bomber had dropped its load.

Jamming Japanese radar initially presented some technical problems due to the unfamiliar characteristics of their equipment. Unlike German radar, the Japanese sets operated on such low frequencies that they were almost invulnerable to the electronic deception of chaff, which had been so effective in Europe. The reason for this was that the tin-foil strips were nowhere near half the wavelength of the radar to be jammed and so did not produce the desired effect.

To overcome this problem, new strips were devised. Made of

aluminium, they were much longer (30 m × 3 cm) and were called 'ropes' because of their shape. The use of this modified electronic countermeasure considerably reduced American losses during incursions over the various airbases built by the Japanese on the occupied islands, which were defended by radar-controlled anti-aircraft gun batteries.

When the Japanese got their hands on some of the new strips launched by the American aircraft, they immediately took steps to install at their airbases other types of radar which operated on even longer wavelengths. They also installed a large number of powerful radar-controlled searchlights which were positioned near the gun batteries. The American bombers, which had recently begun to attack the airbases only by night to cause difficulties for Japanese defences, now found themselves trapped in a web of light beams which continuously illuminated them in spite of their attempts to jam the radar controlling the beams. As soon as signs of interference showed up on the Japanese radarscope, the operator automatically switched the searchlight control onto radar using a different frequency to ensure that the bomber under fire would be continuously illuminated by the searchlight. Using this system, the Japanese managed to inflict high losses on the US Army Air Force in the Pacific: over 80 per cent of the Boeing B-29 Superfortress bombers shot down by Japanese anti-aircraft fire can be attributed to the radar-searchlight-gun system.

Nevertheless, electronic countermeasures did influence the outcome of events in the Pacific to some extent. In the final analysis, the drop in American aircraft losses must be attributed to the large number of jammers carried on board their aircraft (some B-29s carried as many as sixteen), together with the simultaneous use of automatically-launched 'ropes' of varying lengths. Electronic warfare also played a major part in attacks on ship convoys and in amphibious operations.

As we have seen, one of the biggest problems facing the Japanese was to keep sea-lanes open between the mother country and all the occupied islands which were now their territory. When the Japanese entered the war on 7 December 1941, they had a merchant navy of about six million tons but, by the middle of 1943, they had already lost two million tons, which could not be replaced due to the limited capacity of their shipyards. As the Japanese-occupied territories expanded, it became more and more evident that their merchant navy was unable to meet the growing need for long-distance supply transportation to the various islands.

Knowing this, the Americans naturally set about systematically sinking as many Japanese merchant ships as possible by submarine. In an effort to stop this happening, the Japanese equipped their merchant ships with a radar which would give them early warning of the presence of an enemy submarine. However, the Americans countered this by equipping their submarines with RWR and the submarines were, therefore, able to detect the enemy before they were themselves detected. The result was just the opposite of what the Japanese had hoped for, because the American submarines, on intercepting emissions from an enemy merchant ship, were able to home in on it and sink it.

Naturally, the RWR on board American submarines was equally effective against Japanese warships and, particularly, submarines. During the memorable Battle of Leyte Gulf, an American submarine managed to detect three enemy submarines by EW and sink them all.

Two other episodes had important consequences for the war in the Pacfic. One was the Battle of Midway, which marked a turning-point in the war between the United States and Japan.

The Japanese attack on Pearl Harbor, the tragic outcome of which can be largely attributed to serious shortcomings in American electronic organisation, had brought the US Navy to its knees. Consequently, on the eve of the great naval air Battle of Midway, Admiral Nimitz, Commander-in-Chief Pacific Fleet (CINCPAC), found himself with only three aircraft carriers and no battleships at his disposal. On the other side, Admiral Yamamoto, Commander-in-Chief of the Japanese fleet, was in possession of five aircraft carriers and eleven battleships. Nimitz, however, had something which Yamamoto did not have, and this turned out to be of crucial importance.

As a direct result of the Pearl Harbor disaster, the Americans had set up an electronic surveillance network unequalled in the world. All enemy transmissions, both meaningful (such as radio communications) and meaningless (such as radar emissions) were picked up night and day by aircraft, ships and ground stations. All intercepted signals were channelled to a bunker on the island of Oahu where they were anlaysed by code-breakers and electronic experts.

Among the many achievements of this exceptional electronic warfare centre was the cracking of the secret Japanese cipher system and the detection of periodical changes in all enemy coding.

On 20 May 1942, a few weeks before the Battle of Midway, Yamamoto transmitted a coded message to his naval high commands in which he informed them of his plans for the next military operation, Plan 'MO'. By one of those curious twists of fate which turn out to be of crucial importance, the message was mistakenly transmitted in the old code, which the Americans had already cracked, and not in the new one which would have been more difficult to interpret.

After a week's work, the American code-breakers at the Oahu centre were able to understand the text of the top secret Japanese message. Nimitz was duly informed that Yamamoto had decided to attack 'A.F.', probably on 3 June, and had organised a fake attack in the Aleutians to divert the Americans from the site of the main attack at 'A.F.'. The problem now was to find out what locality was indicated by the letters 'A.F.'! How they did this was another masterpiece of American electronic espionage.

Through an accurate analysis of Japanese radio communications, the Americans arrived at the conclusion that the site of the attack must be the island of Midway. An ingenious scheme was devised to confirm this theory.The US forces on Midway transmitted an easily decipherable coded message to headquarters informing them that their water-distillation plant had broken down. The Japanese fell into the trap and, a few days later, Admiral Yamamoto transmitted a message stating that 'A.F.' was short of water due to a break-down of their water-distiller!

Admiral Nimitz now knew where to go and wait for the enemy. He gave orders for the immediate preparation of his three aircraft carriers, *Hornet*, *Yorktown* and *Enterprise*, and set course for Midway. As the two fleets converged on the island, American carrier-borne aircraft made a series of devastating attacks, sinking the Japanese aircraft carriers one by one and forcing the invasion to be cancelled. This American victory had extremely important consequences for the outcome of the war.

The other episode, made possible by the superb organisation of American electronic warfare, was that in which Admiral Yamamoto himself was the target.

In April 1943, the Commander-in-Chief of the re-united Japanese fleets decided to visit his advanced bases to follow the Guadalcanal operations and to inspect defences. On 13 April, the Commander of the Eighth Japanese Fleet transmitted a message to other commands concerned regarding the admiral's planned itinerary. The message stated that Admiral Yamamoto would leave Rabaul on 18 April at 06.00 hours on board a light bomber escorted by six fighters bound for the island of Bougainville at the southeast tip of the Solomon archipelago, where he would inspect the bases at Ballale and Shortland. Arrival at Ballale was scheduled for 08.00 hours on the same day.

This message was intercepted by the American radio stations which were on duty night and day, listening to and recording all enemy electromagnetic emissions. It was then sent to the decoding department where it was promptly deciphered.

On the morning of 18 April, eight USAAF Lockheed P-38 Lightning fighters took-off from Henderson Field on Guadalcanal and waited for the Japanese admiral's aircraft 35 miles to the north of Ballale. When it arrived, they shot it down. Yamamoto was killed.

Thanks to electronic warfare, an American pilot was able to eliminate from the Pacific scene the man who had master-minded the attack on Pearl Harbor, the highly intelligent and greatly revered Admiral Yamamoto. His loss was deeply felt by the whole Japanese Navy.

However, the greatest contribution made by electronic warfare to the Pacific war was in the amphibious landings which took the Americans from Guadalcanal right into the heart of Japan. It was a continuous, though almost unacknowledged, contribution, both prior to and during each operation.

As soon as the Japanese occupied an island, they immediately set up all sorts of early warning and fire-control radars. The American electronic warfare units had to locate all these radars from the Solomon islands to as far afield as the coasts of China and then, to reduce the loss of life during the crucial phases of the operations, neutralise fire-control radars in the areas designated for landing operations.

During the invasion of the Marshall islands in the central Pacific, ships equipped for electronic warfare intercepted the early warning radar installed by the Japanese on one of the islands to warn the local forces of the approach of American ships or aircraft. After studying the technical parameters of the radar, suitable tactics were devised. The information gained from the radar installed in the Marshall islands proved extremely valuable for the US Navy when they attacked the island of Palau a few months later. They were able to install on board their ships jammers tuned accurately to the frequencies of the local radars.

Even more extensive use was made of electronic warfare tactics during the American invasion of the Marianas islands. Prior to the invasion, the Americans carried out thorough electronic reconnaissance of the radar systems operating in the area. These efforts were well worthwhile as they discovered a 'hole' in Japanese radar cover which allowed the invading forces to land undetected by enemy radar.

The importance of electronic measures was again shown in the course of operations in the Philippines. Prior to operations in the Gulf of Leyte, the Americans discovered two radar stations; one was installed in the Gulf of Leyte itself and the other was on the island of Mindanão. These guarded the means of access to the respective beaches and could compromise the success of the invasions. They were therefore attacked and destroyed in order to facilitate landing operations.

Another important event in electronic warfare took place during the famous and dramatic invasion of the island of Iwo Jima. While a US cruiser squadron was moving in to bombard the island, electronic warfare operators noticed that the Japanese had a certain number of fire-control radar sets on the island. Again, it was possible to analyse the characteristic parameters of these instruments and transmit the information to the escort ships. The latter then turned on their jammers to prevent the Japanese from using their radar to aim their coastal batteries at the American landing forces.

12

Telecommunications and Electronic Warfare

Throughout the whole of World War Two, the protagonists const-antly jammed their opponents' radio broadcasts in order to hinder the spread of propaganda by this means. Many people noticed a great deal of interference while tuning their radio sets and sometimes trans-missions were completely drowned by metallic noises, the chiming of bells and so on.

Military communications by radio were also jammed, although to a lesser degree, to prevent the enemy from making effective use of their radio sets. One of the first cases of such jamming took place in November 1941 when the British Eighth Army was preparing a large-scale offensive against the Axis troops on the Libyan Front to regain their lost positions.

During the previous daring operations of General Rommel's armoured columns, the British had noticed that the success of the Germans was partly due to well-organised radio communication between command and the tanks. The British considered that by disrupting these communications, they would be able to paralyse movements of the enemy armour forces. Therefore, a number of rudimentary 50-watt frequency-modulated radio transmitters were installed on Wellington bombers. These transmitted the noise of the aircrafts' engines, producing a deafening, chaotic noise, on the same frequency as that used by the Germans. Initially, the jamming caused great confusion among the German armoured columns but, as soon as they identified the source of the interference, they sent out Bf 109 fighters to shoot down the Wellingtons. This was an easy task as the Wellingtons were slow and were not provided with an adequate escort.

As we have seen in the operations in the Atlantic and Pacific Oceans, one of the most fruitful activities of electronic warfare during World War Two was the interception of enemy radio communications. This activity was carried out by the warring nations not only with the aim of gaining useful information from the decoded messages, but also for the purpose of discovering espionage networks based in their own territory.

An interesting case involving the latter activity was a German operation intended to locate a clandestine Russian radio station operating in German occupied territory.

In 1941, the German military intelligence service, the Abwehr, intercepted at least 500 coded messages which they had been unable to decipher. The Abwehr realised that there was a Soviet espionage network operating in western Europe. Nazi leaders in Berlin were infuriated by their inability to get their hands on this spy network which, being so well-equipped with short-wave radio sets and accessory electronic devices, had been nicknamed the Red Orchestra (*Rote Kapelle*). It was extremely humiliating for them to know that messages containing military information were being transmitted to the Russian military command from inside German territory, but all their efforts to ferret out that den of Russian spies had so far been in vain.

Direction-finders then in existence were not sophisticated enough to give an immediate and accurate fix on the clandestine radio station which, moreover, was continually changing location. It was like a fox-hunt between the clandestine station and the 'Peilung', the German direction-finders, to which service technical improvements were continually being made.

The clandestine station transmitted continuously for four to five hours each night. The Germans systematically intercepted the transmission and, using their DF, calculated the bearings of the station. But, every time, the station was transferred to another locality before it was located by the Germans. Eventually, they managed to establish that the main transmitting station of the Red Orchestra was in a Belgian city. Germany's most skilled DF operators were sent to the city to try to discover the exact location of the transmitting station.

The Russian spies had stayed in one place for too long and this error proved fatal for them. On the night of 13 December 1941, the building used by the clandestine station was located by the expert German DF operators. The spies were caught red-handed by German soldiers who had entered the building undetected, wearing thick socks over their boots to muffle the sound of their footsteps.

It is well known that systematic interception of enemy radio transmissions was an activity that the British, more than other nations, had been engaged in for some time and in which they had gained considerable experience. Immediately after World War One, they had set up clandestine receiving stations all over the world to intercept communications of potential enemy states. All the intercepted messages were analysed and, if possible, decoded with the aim of gaining information which might be useful to them for political or military purposes.

In those days, the decoding of a message was still an activity whose success depended entirely on the skill and intelligence of people who were experts in the field. The encoding of messages was also done by human hand and the secret files containing the codes were kept under lock and key and guarded night and day.

The coding of a message was a long job which involved 'translating' the plain text into a series of numbers and letters which were often doubled-coded to make the enemy's task of decoding even more difficult. Obviously, the greater the number of alphabetical and numerical combinations used, the more difficult the code was to break.

Just before the outbreak of World War Two, the Germans built a special message-coding machine, called 'Enigma', which was able to come up with a far greater number of combinations than had been possible in the previous manual coding procedures. It was an electrical machine with a keyboard to be used by the operator and was able to produce over four million combinations. The German high commands were convinced that the problems of coding would be entirely solved by this machine as it would make the enemy's task of decoding all but impossible.

However, due to a series of events initiated by Richard Lewinski, a Polish Jew who had worked as an engineer and mathematician in the Berlin factory where 'Enigma' had been built, this was not to be. One day in 1938, Lewinski walked into the intelligence service headquarters in Warsaw and offered to sell information regarding 'Enigma' for the sum of £10,000 Sterling and a passport which would allow him and his family to emigrate to England, at that time an ally of Poland.

The Polish and British secret services could not believe their luck, especially as it seemed extremely likely that they would shortly enter into war with Germany. However, the mere possession of the machine was not enough to rapidly decipher the messages, as the Germans changed the 'code-keys' every day. To overcome this problem, the British decided to build a new machine capable of performing all the necessary operations to rapidly find the current 'keys'. These would then be fed into the decoding machine which would give out the enemy message in plain text. About thirty mathematicians set about calculating all the numerical and alphabetical combinations of the 'Enigma' coding-machine, which they had by now been able to construct with Lewinski's help. When this had been done, the calculated combinations were electronically stored in the enormous decoding machine which was thus able to produce, by electromechanical means, the right key to decode 'Enigma' messages.

This system, which was called 'Ultra'[1] was a kind of rudimentary electronic calculator, although the electronic technology used was a far cry from that of modern computers.

When World War Two broke out, the British found themselves in a particularly advantageous position, thanks to 'Ultra' and their world-wide network of intercept stations. They were able to intercept orders from Germany army, navy and airforce commands to their respective forces, communications among themselves and even orders given by Hitler himself before every important military operation.

This incredible 'scoop' by the British secret service had extremely important consequences, especially in the first years of the war. British foreknowledge of enemy operations, provided by 'Ultra', regarding strategy, troop alignments and movements of enemy forces greatly influenced the outcome of many battles.

To give an example, the British victory in the Battle of Cape Matapan on 28 March 1941 can almost certainly be attributed to British decoding of messages sent by the German high command to their Air Corps (X CAT) in Italy a few days before the Italian fleet, which it was to escort, left Taranto.

As we have seen earlier, the Italian navy had planned a surprise attack on British convoy ships in the east Mediterranean using one battleship, four heavy cruisers and six destroyers. The success of this operation depended on the element of surprise and the Italians did all they could to keep their plans secret. However, they asked their allies, the Germans, to help them by sending X Air Corps fighters to protect their fleet. 'Enigma' coded messages relating to this matter trans-mitted by the Germans were picked up and deciphered by the British 'Ultra'. In this way, the British were informed of all the main elements relating to the mission of the Italian naval squadron: date, time, ships employed, air support and so on.

In the light of this information, Admiral Cunningham, ordered the British naval squadron, anchored in Alexandria, Egypt at the time, to get ready for an immediate departure. In order to deceive Italian spies in the port of Alexandria, the British admiral went ashore dressed in civvies and carrying his golf clubs. Under cover of darkness, he secretly returned on board and the fleet put to sea.

'Ultra' also facilitated the destruction of numerous Italian convoy ships bound for North African ports. The British deciphered messages from the German high command to General Rommel and to their Air Corps in Italy regarding departures and arrivals of supplies sent by sea to the Afrika Korps, providing information such as the departure and arrival times of the convoys, ports of departure and destination and the route the ships would take. In this way, the British were kept constantly informed of the departure of enemy convoy ships and could promptly send out units to attack them. Moreover, systematic aerial photo-reconnaissance of Italian ports constituted a great, though little known, advantage for the British as it provided them with further information regarding the convoy ships escort cargo and so on. Another factor in the success of the British fleet in the Mediterranean was the possession of radar which enabled them to fire at night.

On the occasion of the occupation of the island of Crete by German paratroopers in 1941, the British were greatly helped by information gained by the interception and deciphering of messages transmitted by the *Luftwaffe* command to their participating units. Although the Germans succeeded in occupying the island, their losses were extremely high as the British deployed their troops in the precise area where the German paratroopers landed.

Many other Anglo-American successes of World War Two have been attributed to the diabolical machine, 'Ultra': the Battle of Britain, the Battle of El Alamein in North Africa and the invasion of Normandy, to name a few. It is difficult to judge exactly how far 'Ultra' contributed to the success of these operations but there is no doubt that it provided the British with extremely valuable information which must have swayed the course of many conflicts. Perhaps the Italian navy and merchant marine paid the highest price as a result of the 'Ultra' decoding system which provided the British with so much valuable information about their activities.

Knowledge of the enemy's activities and intentions gained by interception and deciphering of communications, together with adequate defence of one's own communications, has alway been a major factor in warfare. Considering the great progress which has been made in the field of electronics for military uses and the growing need for command and control of the armed forces, it has become an absolute necessity to protect communications not only from decoding but also from electronic countermeasures (interception, jamming and deception). Protection of communications is, in fact, a top defence priority in every country today and is considered to be just as important as the acquisition of weapons, the training of forces and all the other major components of modern warfare.

13

The Cold War, Korea and Electronic Rearmament

When World War Two came to an end, both the Americans and the British rapidly demobilised their war machines and EW equipment fell into disuse. Some of it deteriorated due to lack of maintenance and some was even sold to army surplus dealers. Electronic counter-measures fell into oblivion and most of the people who had gained experience in this field during the war disappeared from the scene or moved on to better paid jobs in the electronics industry. Radar, on the other hand, made continuous progress as it had become an indispensable navigational aid for ships and aircraft, above all at night or in conditions of poor visibility.

Unlike Great Britain and the United States, Russia, the other victorious Great Power, was not so quick to demobilise and Soviet forces continued to dominate the scene in Europe and Asia. Using the skill and knowledge of hundreds of German scientists who had been captured in the occupied territories, the Russians carried out extensive research in the field of electronics for military use and began to build electronically-guided missiles.

During World War Two, the Russians, like the Germans, had used their air force almost exclusively to provide tactical support for ground forces and had not, therefore, built large aircraft, like the British and American four-engine bombers, specifically designed for strategic bombing. In the aftermath of the war, they decided to remedy this deficiency in their arsenal by producing hundreds of B-29-type bombers, copied from an American Boeing B-29 Superfortress strategic bomber which had fallen into Russian hands after a forced landing in Siberia.

Meanwhile, as a result of unclear and contentious areas in the peace treaties, disagreements soon arose between the Western Powers and the Soviet Union.

During the early postwar period, it was the atomic bomb, then only in the hands of the Americans, that prevented a new outbreak of war; the threat of an atomic reprisal was a sufficiently strong deterrent to prevent the Russians from taking military action. To give an example, it was the atomic deterrent, as it was called in those days, that prevented war from breaking out when the Russians began their blockade of West Berlin in 1948. The important ex-capital, marooned in Russian-occupied East Germany, and divided into British, American, French, and Russian sectors, with two million citizens living in the Western sectors, was brought to its knees when the Russians refused to allow supplies to be transported by road through East German territory. When the Americans, British and French decided to set up the famous Berlin airlift between West Germany and Berlin, the Russians could easily have occupied the western sectors of Berlin. Their decision not to do this was due to their fear of an atomic reprisal by the Americans, against which they had no defence.

The blockade of Berlin ended in May 1949. It was a moral victory for the western world but also marked the beginning of what came to be called the Cold War between the Soviets and the Western Powers. The Cold War continued for a considerable time and was characterised by brief periods of open hostility and an atmosphere of reciprocal suspicion which resulted in the formation of the two major formal alliances, the North Atlantic Treaty Organisation (NATO) and the Warsaw Pact.

Electronic jamming of communications became an extremely important strategic component of the Cold War. The first act of electronic warfare, in this context, was Russian jamming of programmes transmitted by the Voice of America (VOA) and the British Broadcasting Corporation (BBC) which, transmitted in the Russian language, were aimed at the countries of Eastern Europe behind the so-called 'Iron Curtain'.

When American and British diplomats protested to Moscow and the United Nations that such action was unjustified in peacetime, the Russians replied that the VOA and BBC transmissions constituted an act of psychological warfare against which the Soviet Union had the right to defend itself by paralysing the enemy broadcasting stations.

Russian jamming of western broadcasts went on for many years in spite of the tremendous expense such activity involved. The VOA alone had eighty-five broadcasting stations in Europe and North Africa and employed sixteen different frequencies on both medium and short wavelengths. According to an estimate made in those years, the Russians had something like 1500 jammer-transmitting stations, 800 of which were in Russia and 700 in the satellite countries.

Jammers were designed and built *ad hoc* and were controlled by an extremely efficient interception network. As soon as the VOA changed frequency to avoid interference, Soviet receivers immediately pinpointed the new frequency and continued their jamming. The Russians were so well-organised that the times of their jamming coincided almost exactly with those of the VOA and BBC trans-missions. Although the Americans often managed to avoid being jammed by the Soviets, the latter went on with this activity right up to September 1959 when the Soviet leader, Nikita Krushchev, made an official visit to the United States.

This type of electronic warfare was not confined to Europe. The Chinese, under the leadership of Mao Tse Tung, soon learnt the art of electronic 'interference'.

According to clauses contained in the peace treaties, the Americans had the right of access to Chinese seaports. During the famous 'Long March' to the eastern and southern regions of China, led by Mao himself, the US Seventh Fleet, deployed in the Pacific, did all it could to protect these rights. A few months before the march, a ship specially equipped for communications had been stationed in the Chinese port of Tsingtao to ensure that radio communications could be made between US ships and naval high commands on Guam and other Pacific islands.

One day, however, American radio communications stopped functioning and a strange interference was constantly present throughout the whole network. Suspecting that they were being jammed, the Americans organized an electronic reconnaissance mission, using a small ship equipped with direction-finders, to locate the source of interference. This was rapidly done and the Chinese transmitter which had been causing the interference was promptly put out of action by the US Marines.

It was in this uneasy political-military atmosphere that war broke out between North and South Korea in 1950.

When Roosevelt, Churchill and Chiang Kai-shek met at Cairo in 1943 to decide the future of Japanese-occupied territories in the Far East, it was decided that the Korean peninsula would become an independent free state after the war. However, shortly after the Japanese withdrawal, the Russians occupied the northern part of the peninsula and the Americans occupied the south. Two separate Korean states were thus established. The theoretical border between the states along the 38th parallel soon became the major bone of contention in the growing struggle between the Russians and the Americans all over the world.

Relations between the two new states, communist North Korea and non-communist South Korea, became more and more strained until, on 25 June 1950, North Korean forces crossed the 38th parallel, invading South Korean territory. The United Nations demanded that

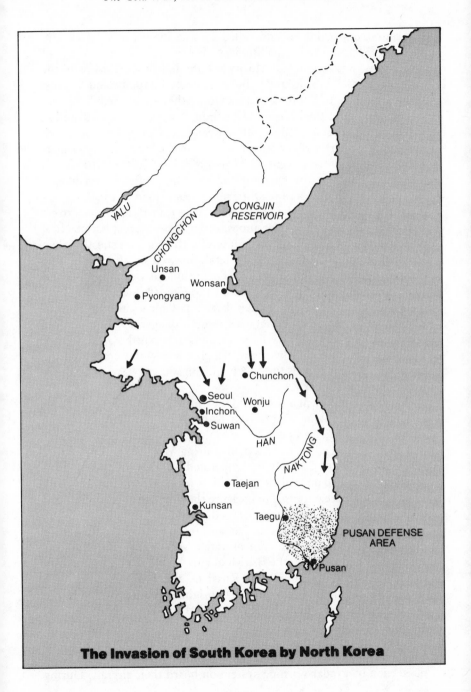

The Invasion of South Korea by North Korea

the aggressors withdrew and called for the intervention of all UN member states. An Expeditionary Force consisting mainly of Americans, was assembled. Meanwhile the North Koreans, with the support of the USSR and China, advanced rapidly southwards, occupying most of South Korea including the capital, Seoul.

Five days after the invasion had begun, American aircraft based in Japan came to the aid of the South Korean forces, providing them with air support. Shortly afterwards, troops from the USA and other non-communist countries went into action in support of the South Koreans. This was the beginning of the long, difficult and bloody Korean War which lasted for three long years, 1950 to 1953.

In the first few months of the war, American B-29 Superfortress bombers were able to operate almost unhindered against both tactical and strategic targets, but the situation changed drastically when Russian-supplied MiG-15 jet fighters appeared on the scene. The Russian fighters had the advantage of being able to use the airbases and long-range radar stations situated on the Chinese side of the River Yalu, which marked the border between China and the Korean peninsula. It therefore became extremely dangerous for the large American bombers to fly daylight missions over North Korea, so they decided to operate only by night, which considerably improved the situation for some time as the North Koreans did not have an adequate night-sighting system.

The only equipment available to the North Koreans were old radar sets from World War Two, given to them by the Russians or the Chinese, and these had very limited range. Two basic types of radar had been developed by the Russians during World War Two: Rus I (Dumbo) and Rus II. They both operated on a low frequency band and were mounted on trailers pulled by trucks and sometimes even by horses. They were only able to measure the distance to an aircraft and give an approximate indication of its direction.

However, the North Koreans had also received from the Russians a number of Mark II fire-control radar sets which the Russians themselves had got from the British during World War Two through the lend-lease programme. The Russians also provided them with several SJ search radar sets which had originally been provided by the United States and which the Russians had subsequently copied and mass-produced.

Although the Americans themselves were ill-prepared for electronic warfare, they were lucky enough to find themselves already in possession of a radar warning system on board their aircraft. During air raids over North Korea, American pilots noticed that, just before they came under fire from enemy anti-aircraft guns, the signal light on their instrument landing system began to flash, thus warning them of the imminent danger. The reason for this was that the Rus II search radar sets used by the North Koreans operated on a frequency (72 Mc/s) very close to that used by the American instrument landing system (75 Mc/s). By this fortunate coincidence, the signal light on the flight panel of the American aircraft, besides telling the pilot that it was time to start coming down for landing, also warned him, while flying over North Korea, that he had been discovered by enemy radar. This warning gave the pilot sufficient time to enable him to take appropriate evasive action.

This piece of luck was short-lived, however, as the North Koreans soon acquired a new radar system, given to them by the Chinese, which used a much higher frequency in the X-band i.e. between 8000 and and 12,000 MHz). As soon as their improvised RWR ceased to work, American losses increased considerably, since they now had no way of knowing that they were within range of Korean anti-aircraft artillery (AAA) which, in the meantime, had been considerably strengthened.

The Americans also noticed that enemy anti-aircraft fire had become more accurate even in conditions of poor visibility. They became convinced that the North Koreans must have got hold of a new type of radar. They therefore hastily retrieved equipment which they had disposed of at the end of World War Two. They even had to buy some of it back from army surplus dealers! As soon as this equipment was in working order, it was sent to the Far East and installed on aircraft operating in North Korea. These old receivers were not capable of picking up the emissions of the new enemy radar, thus confirming the Americans' suspicion that this new radar operated on a much higher frequency.

As soon as the Americans installed new receivers and acquired detailed information regarding the characteristics of the new enemy radar (frequency, pulse width and pulse repetition, etc.), World War Two jammers were modified and installed on old North American B-25J Mitchell bombers which were then given the task of protecting the B-29 bombers during their air raids over Korea. The old system of jamming enemy radar by means of tin-foil strips, called chaff by the Americans, was also resurrected.

Confronted by these electronic warfare devices Korean radar was no longer able to pick-up the enemy bombers or guide their searchlights and AAA onto them: American losses once again began to decrease.

While all this was happening in the skies above Korea, the war on the ground progressed erratically, first in favour of one side, then the other. In 1950, a few months after the war broke out, UN Forces had managed to regain all the territory constituting the Republic of South Korea, thanks to the landing at Inchon, on the west coast of the Korean peninsula, behind enemy lines. Fighter-bombers from four aircraft carriers and the 5th Air Force, 250 ships, 70,000 men, including a division of US Marines, took part in this landing which was made difficult by the geographical features of the area, including the steep cliffs which had to be scaled, very powerful tidal flows and frequent typhoons.

It had been clear from the start that the success of the operation depended on the choice of the date for the landing and on the coordination between the landing troops and the supporting air and naval forces. The date was fixed for 15 September and, a few days before, continual photographic and electronic reconnaissance flights were made to locate the radar stations which had to be put out of action. However, thirty-six hours before the time set for the landing, a violent typhoon hit Japan and Korea, causing serious difficulties for the smaller ships and landing craft, as well as for air operations. The landing, nevertheless, took place according to plan, following a massive air and naval bombardment. Thanks to perfect planning and the support of hundreds of F4U Corsair and AD-1 Skyraider close support aircraft, the amphibious assault was a complete success and, after seven hours of fighting, all the objectives of the operation had been achieved.

When the UN Forces, having crossed the famous 38th parallel, drew near to the border of Red China, the latter started to send an ever increasing number of volunteers to help the North Koreans. The Americans and their allies were forced to retreat southwards. It was a bloody struggle with disappointing results. In spite of the thousands of aircraft from US Navy and Royal Navy aircraft carriers and the US Marine Corps, air support did not achieve great results due to the lack of important military targets on Korean territory.

UN losses amounted to over 1300 aircraft and, according to later calculations made by American experts, this figure would have been trebled if the electronic actions mentioned above had not been carried out.

The war ended on 22 July 1953, leaving things more or less as they had been before. The 38th parallel was once again the theoretical border dividing the two Korean republics which where governed as before but which were now even more poverty-stricken. The devastation of the entire peninsula left a toll of almost two million dead, wounded or missing, among whom were Koreans, Chinese, Americans and UN soldiers.

The Korean war provided yet another demonstration of how electronic warfare can help to cut losses, especially in the air. Consequently, immediately after this war, there was a great 'electronic rearmament'. All the major world powers dedicated their efforts to producing new types of equipment to enable their bombers to penetrate enemy air space without being detected by radar and fired at by electronically-guided weapons systems.

Shortly after the Korean War, Russia exploded her first atomic bomb and the two great world powers, the United States and Russia, began to realise what disastrous consequences would ensue if one of them were to launch an atomic attack against the other. The possession of the means to cause such tremendous devastation made both countries extremely wary of each other.

The advent of the atomic bomb and, subsequently, the hydrogen bomb (H-bomb) gave war a new aspect and new theories concerning strategy were proclaimed, such as NATO's 'massive response'—a devastating nuclear reprisal against any attacker.

The United States began construction of large, so-called 'strategic' bombers which, in a single air raid, could cause unimaginable destruction by dropping their load of atomic bombs. The first atomic bombs dropped on Hiroshima and Nagasaki at the end of World War Two had been carried and dropped by Boeing B-29 Superfortress bombers and, in the early years of the Cold War it was the B-29 which was used by the American Strategic Air Command (SAC) to carry the lethal weapons.

In the 1950s, the B-29 was replaced by the new B-50 bomber and, later, by the gigantic Convair B-36. In addition to six piston engines, they also had four jet engines under the wings, giving them a ceiling of nearly 50,000 feet, a range of 10,000 nautical miles and a speed of approximately 430 mph. At the end of the 1950s, the first jet bomber, the Boeing B-47 went into service. These were in turn replaced by the famous Boeing B-52 Stratofortress which could carry a huge bombload, flew at an altitude of nearly 55,000 feet, at a speed of over 630 mph and had a range of 12,500 miles. On the Russian side, stategic bombing by the late 1950s was entrusted to the Tupolev Tu-16 (code-named *Badger* by NATO) and the Tu-20 (*Bear*).

Meanwhile, Great Britain and France had also developed the atomic bomb and they too began to build aircraft suitable for delivering this new weapon. Since their sphere of action was limited to Europe, these aircraft were not so impressive as the American bombers. The British built a series of medium-range bombers, the Vickers 'Valiant', the Avro 'Vulcan' and Handley-Page 'Victor', while, for their *force de frappe*, the French built the Dassault-Mirage IV-A, which entered service in 1964.

At the same time, construction of huge, complex air defence radar chains was begun. These had the function of providing early warning in the event of an enemy air attack. The United States built three such radar chains to protect their territory. One ran along the northern border of the United States, another stretched across the central part of Canada and the last, the most advanced, ran from Alaska to Greenland. All the various radar stations were connected by a complex cable and radio communications network which was in operation twenty-four hours a day.

As well as guarding the Arctic, the United States built a network of radar stations along the Pacific and Atlantic coasts. A number of these radar stations were installed on special platforms in the ocean several miles from the coast. Finally, to ensure detection of hostile aircraft coming from either the east or west, a very advanced surveillance service was put into operation, using four-engine Lockheed C-121 Constellation aircraft equipped with long-range radar and other special devices for long-range detection.

In Europe, the NATO countries also began construction of a gigantic radar chain which was to stretch from Norway to Turkey.

Between 1947 and 1949, before the Korean war, the Russians had also dedicated their efforts to building an air defence radar chain, guided missiles and a force of heavy bombers. All this was done in great secrecy and very little information managed to leak out even through the usual channels.

In order to fly past these radar chains without being detected, it was, of course, indispensable for bombers to carry electronic equipment capable of neutralising the radar. The Western Powers, who had learnt, often at their own expense, how important it is to know the charactertistics of the enemy's radar systems, realised that they knew practically nothing about Russian radar and were, therefore, unable to develop appropriate electronic counter-measures. The international political situation was extremely delicate and there was the constant danger that serious tension between East and West, such as that caused by the Berlin crisis, could transform the Cold War into a Hot War!

Aware of this possibility, western nations, particularly the United States, set about gaining information concerning Russian radar by conducting intensive electronic intelligence (ELINT) missions; the Russians, of course, retaliated by doing the same thing.

From 1949 onwards, the gathering of electronic data regarding a potential enemy became a top priority activity involving specially-equipped aircraft and naval units and disguised ground reception stations. Real electronic espionage had begun and was carried out along the borders of potentially hostile countries, at the limits of their territorial waters or in the skies above the countries themselves.

The instruments used in these ELINT missions were mainly the following:

receivers and interceptors of various types to pick up the electromagnetic emissions of the radars of potentially hostile countries analysers to examine the intercepted emissions and ascertain their main characteristics;

direction-finders to establish the direction of arrival (DoA) of the emissions themselves and to pinpoint the location of the emitting station;

a series of recorders of various types to store the information for further, more detailed analysis.

The aim of such activity was to find out the enemy's 'electronic order of battle', in other words, the deployment of radar stations in all areas under observation, in order to devise appropriate counter-measures to be used if or when they were called for.

Knowledge of the electromagnetic situation of a potential enemy and of variations of his situation has many important side-effects. One is that it permits experimental launchings of new, intercontinental ballistic missiles by the potential enemy to be monitored because such test launches entail a series of checks on radar guidance systems, tracking radar[1], radio communications equipment and the telemetric instruments of the missile itself. Interception and analysis of these electromagnetic emissions can reveal whether the experimental launching of a new missile is, in fact, being carried out, while the electronic information gathered reveals whether new missiles have been deployed in certain zones and what technological progress the enemy has made in the design and construction of electronic systems. This electronic intelligence also permits a fairly accurate picture of the enemy's defences and sometimes even his political-military intentions to be built up. In short, one can obtain a general picture of what is commonly called the 'threat', which is merely the sum total of the capabilities and intentions of the opponent, in an arc of 360 degrees around one's own country. From a strictly electronic point of view, the threat comprises all the actual or potential enemy's methods of utilising electromagnetic energy for the guidance of weapons, command and control of forces, and surveillance of the theatre of operation.

These electronic reconnaissance missions, called 'ferret missions', were often quite risky as it was necessary for the aircraft or vessel to penetrate the potential enemy's air space or territorial waters in order to achieve the aims effectively. In fact, it was not merely a case of collecting data concerning radar and radio but was often also a 'challenge' to test their reaction in terms of time and efficiency.

In a typical ferret mission, an electronic reconnaissance aircraft would simulate the role of a bomber penetrating enemy air space. It would fly directly towards the border of the 'enemy' country, often even flying over it. In this case, it was usually detected and tracked by the country's long-range radar system. On no account had the aircraft to attempt to avoid detection because electronic warfare specialists had to record the frequency and PRF (Pulse Repetition Frequency) of the enemy's long-range search radar and locate its position on the map. Immediately after target acquisition by the enemy search or acquisition radar, enemy interceptor aircraft would be launched to intercept the intruder. At this point, the crew of the spy-plane had to measure the electronic parameters of the search radars' emissions and determine how much time had elapsed between target acquisition and the interceptors' taking off. If anti-aircraft batteries were brought into action, the ELINT operators in the spy-plane had to measure the characteristics of the fire control radars too. Sometimes, they even had to take note of the time it took the batteries to fire the first round and, where possible, evaluate the accuracy of their fire.

The highly-specialised crew members who took part in these missions were men of great skill and ability and, above all, tremendous courage; with every mission they flew, they risked both their own lives and the possibility of provoking extremely serious diplomatic incidents.

In spite of this, such missions were an everyday occurrence and no government dreamt of protesting. The governing principle of the time was that of reciprocal action; tit for tat, as is often the case with international relations in peace-time. The only rule was, whatever you do, do it well without getting caught.

The aircraft used in these missions had to have a very long range and had to be able to fly at very high altitudes outside the range of enemy anti-aircraft artillery and at a speed which would make it difficult for interceptors to catch them.

In the early days, the Americans used World War Two bombers, specially-equipped and structurally modified for the purpose of electronic espionage. Among these was the B-24 Liberator, and the US Navy version, the PB4Y2 Privateer. B-29 Superfortress and B-50 bombers, respectively redesignated RB-29 and RB-50 (R standing for reconnaissance), were also used occasionally. Besides the normal crew, these carried a number of electronic operators, each of whom was responsible for surveying a sector of the electromagnetic spectrum. Later, the US Navy twin-engine Lockheed P2V Neptune, a maritime patrol (MP) aircraft, was used in ferret missions. The Neptune was famous in its time as it held the record for the longest flight ever made. Cruising range was a fundamental characteristic to be considered when selecting an aircraft for this type of mission. Ferret aircraft often had to patrol the area under observation for long periods of time before the enemy radar emitted pulses which they could intercept and record. For this reason, transport aircraft, such as the C-47 and the C-18, were also used for these missions. The American's successor to the Neptune was the P-3C Orion, also built by Lockheed, the military version of the Electra turboprop transport.

When the risk of being attacked by enemy fighters became a real possibility, the US Navy introduced the Martin P4M1Q Mercator which was specially designed for this particular activity, had excellent range and four engines, two of which were jets, enabling it to accelerate rapidly and escape attack should an enemy fighter suddenly appear on the horizon.

On the Russian side, ferret missions were carried out by Tupolev Tu-16 *Badger*s which, in the original version first observed in 1953, were long-range bombers each carrying two air-to-surface *Kennel* missiles, later replaced by missiles of a similar type code-named by NATO *Kelt* and *Kipper*. The Tu-16D version was employed solely for ELINT and was easily recognised by the radomes (domes covering radar) on its fuselage. At first, these aircraft operated in the Pacific, carrying out electronic reconnaissance on the US Seventh Fleet and its bases in the Pacific Ocean. Their main base was at Petropavlovsk on the Kamchatka peninsula. Each Tu-16D carried, beside the crew, seven operators and a radar-officer, all specially trained in electronic reconnaissance. The sphere of their reconnaissance was later extended to the Mediterranean and the North Sea.

Soviet ferret missions were, of course, similar to those carried out by the Americans. Every operator had a receiver to intercept electromagnetic signals in a specific sector of the spectrum, a pulse anlayser, a direction-finder to calculate the direction of arrival of the emissions and, finally, a number of special recorders to record them. Each operator had to carefully survey the portion of the spectrum assigned to him, noting down in his log-book all signals intercepted and recording those which seemed particularly interesting. The L-band (1000 – 2000 MHz) and X-band (8000 – 12,000 MHz) frequencies were the most commonly used in this period.

In the course of a typical Soviet ferret mission in the Pacific, the aircraft would take-off from Petropavlovsk and fly towards the assigned zone. The operator in charge of the L-band sector woul begin to intercept the first weak signals coming from American search radar installed in the Aleutian islands, which had the task of detecting potentially hostile aircraft at longe range. The aural signals emitted in the L-band were easily distinguishable in the operator's headphones due to the characteristic tone produced by the PRF.

As the aircraft continued on its course, the operator dealing with the X-band frequencies would begin to hear in his headphones the fast bleep of the fire-control radar which usually operated on this frequency band. This meant that the Soviet aircraft had been picked up by the American radar and was being tracked as a potentially hostile target. If, at this point, the aircraft did not change course and head away from the missile base, it would probably be an easy target for the Nike-Hercules surface-to-air missiles (SAM) which, with the Hawk missiles, were the main air defence weapons then deployed by NATO countries. Therefore, at this point, the Tu-16D headed back to base with its precious reels of magnetic tape on which were recorded American radar signals and radio-telegraphic communications between American command posts, control centres and air bases in the Far East and the Pacific. As soon as the aircraft had landed, this material was sent to the Russan Signal Intelligence Service Centre which was located in a concrete bunker hidden in a forest near Moscow. Here the signals were closely analysed by electronic warfare experts who tried to determine the characteristics of American radars in that area and discover any innovations made.

Soviet aircraft also carried out missions of this kind over the waters of Alaska where the chances of intercepting American radar signals were much greater due to the presence of the long-range search radar chain and numerous military bases, and the substantial air, naval and ground forces stationed in the Alaska area. Russian aircraft sometimes penetrated as far as 50 miles into American territory in this region. During one such mission, which took place some years ago, two Russian aircraft, flying at a speed of over 650 mph and an altitude of about 33,000 feet, stayed over Alaska for about half an hour, but kept out of range of the Nike SAMs. The Americans sent out a patrol of four F-102 interceptors to warn off the intruders which headed back to their base as soon as they sighted the F-102s.

An electronic espionage activity in which the Russians excelled was that involving the leech-like presence of specially equipped ships and aircraft in all the areas where NATO naval units carried out their periodic sea exercises to monitor electronic activity. Russian aircraft often flew right over the NATO naval formations, particularly the aircraft carriers, and the NATO forces had no option but to use fighters to chase them away.

The Russians also employed a number of large, motorized fishing vessels which were usually stationed along the American coasts. Besides carrying fishing nets, these boats also had a large number of receivers and special antennae whose function was obvious. These vessles were often stationed near NATO missile-launching bases, lying in wait for the launching of any new type of missile. That this was their main aim can be easily deduced from the presence of multiple helical antennae, the type of antenna most suited to the interception of electromagnetic emissions from missile guidance or fire control radars.

In April 1960, the Russian spy-trawler *Vega* carried out a lengthy ELINT mission in the waters off Long Island, USA, where the Americans were conducting test launches of the Polaris missile from the first US Navy nuclear submarine *George Washington*.

The Russians also employed large, specially-modified ocean-ographic ships for electronic intelligence. Besides collecting oceano-graphic data during their long missions, they also collected data concerning electronic warfare. Submarines were also used by both the Americans and the Russians for this purpose, although they were totally unsuited to the task, particularly because they had to surface in order to pick up radar signals and, thus, risked being detected by those same radar stations. Nevertheless, electronic espionage carried out by both Russian and American submarines must have been fairly intense because, in 1961, diplomatic protests were made by both sides regarding this activity.

However, no Russian aircraft was ever shot down during the course of an ELINT mission nor were there any serious incidents involving other kinds of platform used by the Russians for such activities. On the other hand, about twenty-six American aircraft were either shot down or forced to land in Russian territory or elsewhere behind the Iron Curtain.

There are two main reasons for this discrepancy. First, Russian aircraft rarely penetrated deep enough into enemy air space to come within range of NATO missiles while American aircraft, on the other hand, often penetrated the air space of the Communist bloc and even flew right across their territory. Secondly, the NATO nations were reluctant to launch their missiles against unidentified aircraft, particularly because, in that period, many pilots from east European communist countries defected to the west with their aircraft and, consequently, it was difficult to know whether the pilot of a military aircraft coming from a communist country was a spy or a defector seeking political asylum.

The Russians also differed from the Americans in their observation of radio/radar silence. Whereas the Americans kept their radars functioning virtually all the time, the Russians nearly always turned off their radar equipment when they detected American spy-planes, thereby denying the Americans the opportunity of intercepting emissions and thus locating the Russian's radar stations. Only when suspicious foreign aircraft penetrated communist air space to the point where they feared an attack, did the Russians switch on their radars. It was precisely for this reason that American pilots on ELINT missions, which involved penetrating communist air space were ordered to simulate a real attack so that the radar operators would indeed switch on their equipment. Only by using this deception could American aircraft intercept and record communist radar and radio emissions. Unfortunately, using this tactic risked provoking the air defences into responding with live weapons.

One of the first incidents which can be attributed to this risky activity was the disappearance, in April 1950, of a US navy PB4Y2 Privateer. This large aircraft, which carried a crew of ten, six of whom were electronic technicians, took off from Wiesbaden in West Germany 8 April 1950. It was officially flight-planned to fly to

Copenhagen, but it is extremely likely that its main task was to carry out an ELINT mission in the Baltic Sea area. At 14.40 hours, it made its last radio transmission over Bremerhaven in West Germany.

According to the Russians, an aircraft, which they identified as a B-29 bomber, was picked up at a distance of about 350 miles from Copenhagen, over Leyeya (Latvia), 7 miles within Russian territory, where it was intercepted by a patrol of Soviet fighters and ordered to land at a Soviet airfield. The Russians maintained that the American aircraft had opened fire on the fighters which had then shot it down.

All evidence seemed to point to the fact that the 'B-29 bomber' was, in fact, the Privateer, and the US government bestowed military decorations on the crew who had sacrificed their lives in the performance of their duty.

Incidents of this kind happened all over the world, from the Baltic Sea to East Germany, from Russia to Czechoslovakia, from the Black Sea to the China Sea, from Korea to Siberia, but many of them have never been brought to light.

To get an idea of the strictness of Russian radio and radar silence, it is enough to look at the occasion of Krushchev's visit to Britain in April 1956. The Soviet Communist Party Secretary, Krushchev, and the Soviet Premier, Bulganin, left a port in the Baltic Sea on 16 April 1956 on board the cruiser *Ordzonikidze* escorted by two destroyers, *Smortryashchy* and *Sovershenny*, bound for Portsmouth, England. Several NATO secret services had set up a network of receivers along the route the Russian ships would take, using naval units, ELINT aircraft and ground intercept stations. However, during the entire voyage, which lasted three days, not a single signal was emitted by the Russian ships.

While the *Ordzonikidze* and her escort were anchored at Portsmouth, Lieutenant-Commander 'Buster' Crabb, a British ex-corvette captain and renowned frogman, disappeared in the waters of the port; his headless, armless body was found only after several days. Rumour had it that he had met his end while trying to collect data regarding the Russian ships' sonar and the operating frequency of their underwater emissions: this rumour has yet to be disproven.

An important instrument for gathering electronic warfare data is formed by networks of ground stations which, positioned opportunely, can intercept a large number of radio communications or radar signals and pinpoint, by triangulation, the positions of the various tranmsitters. Consequently, all world powers, great and small, began to set up, or strengthen, these special networks of receiving stations.

Naturally, this activity was top secret. Nevertheless, it is common knowledge that extremely efficient radar intercept systems were set up along the border between East and West Germany, one by NATO and the other by the countries of the Warsaw Pact. It is also beyond doubt that an excellent piece of interception work was carried out in the Persian Gulf, between 1948 and 1950, by a team of British electronic operators masquerading as archaeologists!

However, the most important SIGINT (Signals Intelligence) centre was set up in Iran. The countries of the west were particularly interested in this zone of the Middle East as the Russians had established a ballistic missile range at Tyuratam, between the Caspian Sea and Lake Aral. In order to follow Russian progress in the field of guided weapons and, at the same time, to acquire information regarding the characteristics and performance of the relative radar-guidance systems, the Americans decided to set up special receiving stations in Iran close to the Russian missile range.

These stations, equipped with the most sensitive and precise instruments the electronics industry could produce, were set up at Kabkan, near Mashhad, in the northern mountains near the Russian border, and at Behshahr on the Caspian Sea. They were in operation continuously and, whenever tests were carried out by the Russians on new missiles, the American operators were able to calculate the missile's trajectory by triangulation, and measure all the parameters of the new radars. In this way, the Americans could devise appropriate ECMs to jam or deceive these radars in the event of war.

During the Cold War period, American interest was not limited to the interception of data regarding intercontinental missiles. They were also interested in the strategy and tactics of Soviet air forces. To acquire information in this field, new and more sophisticated listening posts were set up in England (Chicksands), Germany (Darmstadt and Berlinhof), Italy (Brindisi), Turkey (Karamursel and Trabzon) Crete and in a number of locations in the Pacific. The main task of these stations was to intercept and record all communications between Russian aircraft and between the aircraft and their commands. The aim was to acquire information regarding the performance of the aircraft, and their missiles and radar, as well as about operational procedures employed. Some of these stations had gigantic dish antennae covering 360 degrees which were able to pick up radio signals coming from aircraft thousands of kilometres away.

During the worst period of the Cold War, aircraft were also targets for electronc deception which sometimes led to highly dramatic moments, although such incidents are not widely-known. False radio navigational signals were transmitted to aircraft by dummy stations. Amongst the fake signals were ADF (Automatic Direction-Finding), radio-beacons, TACAN[2] and other navigational aid systems. In Turkey and West Germany, for example, there were several cases of NATO military aircraft being deceived into landing on the wrong side of the Iron Curtain. Operating on the same frequency, Soviet radio-beacons would use the code-names of western radio stations in bordering counties or simply furnish false information regarding the route the aircraft should take to come in for landing. It was reported that a Russian warship, anchored in the port of Alexandria, Egypt, in the Mediterranean, imitated the coded response of a US aircraft carrier's TACAN to an F-4 Phantom aircraft which almost caused a serious accident.

14

Electronic Espionage in Peacetime

The Mysteries of the U-2

In the early months of 1956, a strange aeroplane was sighted in the skies of England, Turkey and some other NATO countries. It became the object of great curiosity among the citizens of these countries, some of whom wrote to their local newspapers to find out what type of aircraft it was and what it was doing there. When interviewed by the press, representatives of the various military air forces invariably gave an evasive response or refused to make any comment whatsoever. An official explanation was finally given by the United States, maintaining that it was a Lockheed U-2 aircraft used for collecting data regarding air turbulence and currents, cosmic rays and the concentration of elements such as ozone and aqueous vapour in the atmosphere.

The Americans did all they could to keep the aircraft hidden from 'unauthorised eyes' but, in spite of their precautions, several people managed to catch a glimpse of it and those who got a close view immediately realised that it must be an aircraft specially designed to carry out top-secret missions. In Russia, where many pilots had seen the aircraft flying at altitudes they could not reach, the U-2 was nicknamed 'the black lady of espionage'.

The aircraft was, in fact, painted all black to make it difficult to sight optically at very high altitudes and its real mission was to fly beyond the Iron Curtain, taking photographs and collecting data concerning electronic warfare. It had been designed in 1950 for the purpose of keeping the governments of the United States and other western powers informed about Soviet missile systems and the electronic characteristics of the radars used to control them.

The US Air Force had not been satisfied with the results of the numerous photographic and electronic reconnaissance flights carried out over Russia in the years 1950–55 using ordinary aircraft; during this period, there had been fifteen 'accidents', in which a total of ten American aircraft had been lost. Consequently, the task of organising reconnaissance over Russia had been assigned to the Central Intelligence Agency (CIA)[1]. Their first step had been to commission the Lockheed Aircraft Corporation to design and build an aircraft suitable for this type of activity.

And so the U-2 was created, a real gem of aeronautic technology. It was a cross between a jet fighter and a glider, with a single turbo-jet engine and high aspect ratio wing with a span of approximately 100 feet. It had a ceiling of over 100,000 feet, a range of over 4500 miles, a top speed of about 500 mph and an endurance of about ten hours. To make it lighter in flight, and, therefore, give it a greater range, it was able to drop-release its landing gear after take-off and land like a glider on two skids.

Aircraft capable of reaching such high altitudes as these had been built by the Americans, the Russians and the British, and perhaps by others too, but these were all experimental aircraft built for the purpose of making or breaking records. They could only fly at such high altitudes for very short periods of time and their ability to manoeuvre was severely limited by the rarefied atmosphere and their narrow wings. Moreover, according to CIA experts no surface-to-air or air-to-air missile then in existence was capable of reaching such incredible heights. The U-2 could, therefore, operate in safety, high in the skies over Russia, with no fear of being attacked by an enemy aircraft or missile.

On several occasions, the Russians had tried to shoot down U-2s with fighters and missiles, but all their attempts had failed. Moreover the U-2 was practically immune to radar detection as it was constructed mainly of plastic and plywood. Only the engine reflected radar waves but this was insufficient for the aircraft to be detected unless its exact position and route were already known. Very few of the U-2's flights over Russia were detected by Soviet air defence as its echo on the radarscopes was barely perceptible, even for the most experienced and expert operators.

The features mentioned above were not the only wonders of the U-2! There were eight fully-automatic cameras on board which could photograph almost any object on the sea or land, in daylight or darkness, in fair weather or foul, from incredible heights; the images produced by these cameras were so clear that, from an altitude of about 80,000 feet, one could distinguish a pedestrian from a cyclist or a man in uniform from one in civilian clothes; from an altitude of about 50,000 feet, one could read the headlines of a newspaper or the billboards posted on the walls of a city; from an altitude of about 30,000 feet, one could even see a nail lying in the road! In less than four hour's flying time, a single U-2 could photograph an area of 780 km by 4300 km; a country the size of Russia could be photographed in the course of a few weeks!

The CIA had also ordered extremely sophisticated electronic equipment to be built for electronic espionage missions over Russia. The normal equipment of a U-2 comprised an intercept receiver capable of picking up all signals coming from Russian radars, a receiver capable of picking up all Russian air defence radio communications, a DF to measure the direction of arrival of all intercepted emissions, a very special magnetic tape-recorder which recorded all intercepted electromagnetic emissions, plus, of course, a radio-compass, an autopilot and a UHF radio.

Everything concerning the U-2 was cloaked in the utmost secrecy and both official documents and aviation journals referred to it as a meteorological reconnaissance aircraft; in fact, in 1957, it was given out that the U-2 had photographed a typhoon in the Caribbean Sea. However, in spite of all the efforts made to keep the real activity of the U-2 a top secret, the curtain of silence began to rise following several incidents involving the aircraft which made people rightly suspicious about the mysterious aeroplane.

Regarding the first three or four incidents, which happened in the United States and Germany, the press merely spoke of an aeroplane which carried out missions classified as 'top secret' but, when a U-2 was obliged to make an emergency landing in a glider field in Japan, a local journalist who happened to be there at the time was able to inspect the aircraft for a full fifteen minutes before soldiers arrived and surrounded the damaged aircraft, pointing their machine-guns threateningly at bystanders to make them go away. The journalist, who was also a pilot, saw the pilot of the U-2 climb out of the damaged aircraft and noticed that he had no stripes on his flying suit and that he was carrying a gun. The journalist put two and two together and came to the conclusion that the aircraft was used, not only for meteorological reconnaissance, but also for espionage purposes.

The US aircraft that took off from the air base at Incirlik, in Turkey, at 06.26 hours on 26 April 1960 to fly a photographic and electronic reconnaissance mission over the heart of the Soviet Union was also a U-2. The pilot was Francis Gary Powers, a 30-year-old former US Air Force captain and unanimously considered to be an excellent pilot and a superb navigator. Powers had logged over 500 hours on the U-2, mainly over Russia, and these missions had become a matter of routine to him; he jokingly referred to them as the 'milk run'. The CIA conducted these flights on a regular basis as it was only in this way that effective and useful results could be obtained: in fact, by flying over and photographing a certain area at regular intervals and

Fansong radar used to guide SAM-2 missiles. (By courtesy of *Defense Electronics*)

A Teledyne-Ryan remotely-piloted vehicle (RPV) with electronic warfare equipment, used for battlefield surveillance and ELINT.

A Soviet *Styx* surface-to-air anti-ship missile being inspected before installation. In the background is an 'OSA-2'-class fast attack craft with missile-launchers.

A Soviet 'OSA-2'-class fast attack craft (missile).

A photograph that sets a turning point in the history of naval warfare: taken from an Israeli missile-launcher boat on the night of 6/7 October 1973, during the battle of Latakia. The rocket-propulsion flame from a *Styx* missile is quite visible, reflected in the water. The missile, launched by a Syrian unit, is directed toward the Israeli unit from where the picture was taken, but the ECM equipment on board succeeded in deceiving the missile off course.

A Soviet P-12 *Spoon Rest* radar, with a range of 270 kms, used in the 1973 Arab-Israeli war for early warning in co-ordination with the SAM-2 system's *Fan Song* radar. (By courtesy of *Defense Electronics*)

The USAF Lockheed SR-71 Blackbird electronic and photographic strategic reconnaissance (SR) aircraft has a speed of Mach 3 (almost 38 miles per minute) and a ceiling of 120,000 feet. On 1 September 1974, an SR-71 covered the distance between New York and London in 1 hour 55 minutes. (By courtesy of *Defense Electronics*)

A Soviet-built SAM-6 *Gainful* system and (inset) the missile-guiding *Straight Flush* radar. ECM equipment carried by the US aircraft used by the Israelis proved useless because the radar emissions were CW (continuous wave) and the US ECM equipment was designed to counter only pulse emissions.

The Soviet-built mobile anti-aircraft gun system ZSU-23-4 *Shilka*, used for the first time in the 1973 Arab-Israeli war. Its *Gun Dish* radar, operating on a higher frequency than those previously used, and its computer enabled the *Shilka* system to fire with great accuracy even while in motion. (By courtesy of *Aviation Week & Space Technology*)

The Soviet infrared homing surface-to-air missile SA-7 can be carried by a soldier, as illustrated, or mounted on a tracked vehicle. (By courtesy of *Defense Electronics*)

Below left: A military vehicle photographed in daylight.

Below: The same vehicle photographed at night using an infrared system. (By courtesy of *Defense Electronics*)

Night vision (starlight) image-intensifier aiming equipment mounted on a standard US Army rifle. (By courtesy of *Defense Electronics*)

US air-to-surface AGM-65 Maverick missile installed on the port inboard weapons pylon of an F-4E of a USAF Tactical Training Wing. Left: TV-guided; Right: IR-guided.

155 mm artillery shell with passive laser guidance nearing, then hitting a surplus US target at the White Sands firing range, New Mexico. The fins deploy in flight to guide and stabilise the shell. The inset shows an inert practice round.

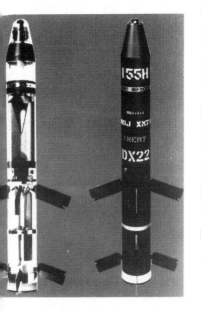

The Soviet spacecraft communications ship *Vladimir Komorov*, equipped for ELINT missions. (By courtesy of *Defense Electronics*)

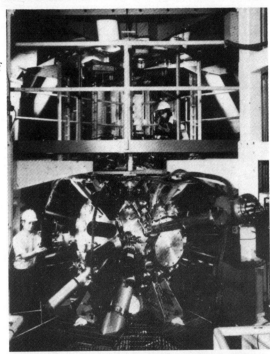

Installation in the Lawrence Livermore Laboratory, California, USA, used for experiments related to high-energy lasers. (By courtesy of *Aviation Week & Space Technology*)

A BAe Nimrod of the Royal Air Force, equipped for electronic surveillance and used for AEW (Airbone Early Warning) in defence of the UK Air Defence Region, (UKADR), alerting interceptors to intruders. The UKADR is the most critical NATO ADR.

The electron accelerator installed at Kurchatov Institute, in the Soviet Union, where experiments relating to charged particles technology are carried out. (By courtesy of *Aviation Week & Space Technology*)

The six camera lenses of a US satellite for photographic and electronic reconnaissance. Up to now, neither the Soviets nor the Americans have ever released pictures of the electronic designs of satellites, for obvious security reasons.

Soviet Cosmos 381 satellite, presented at the Paris Air Show in 1976 as a satellite for ionospheric research.

US Navy Maristat communications satellite launched in 1976.

Soviet reconnaissance satellite: when it orbits over a NATO radar station (left), it intercepts and records the emissions, later retransmitting them as it orbits over a Soviet ground station. (By courtesy of *Defense Electronics*)

The Soviet consulate building in San Francisco. According to United States sources, it contains sophisticated electronic equipment for the interception of electromagnetic transmissions. (By courtesy of *Defense Electronics*)

The Pershing II missile, a derivation of Pershing 1A, has become, as a result of its advanced electronic system and improved mobility, a most accurate weapon, at the moment almost invulnerable to enemy measures and electronic counter-measures. Its first stage of flight is made under inertial guidance, until a target seeking guidance system called RADAG (Radar Area Guidance) takes over. This system operates in the J-band (12 to 18 GHz) to scan the target area and compares this radar display with data collected by reconnaissance aircraft and satellites, and stored in a computer. This system should allow a range almost double that of the Pershing 1A, which means a minimum of 160 km (100 miles) to a maximum of 840 km (525 miles). Its CEP (Circular Error Probability) is approximately 40 metres (130 feet). The missile is mounted on special vehicles (tractors, vehicles with telescopic lift) and can have one or two nuclear warheads, with independent final guidance.

The Soviet helicopter carrier *Moskva*, closely observed by a US Navy Lockheed P-2V Neptune maritime surveillance aircraft. (By courtesy of *Defense Electronics*)

The Soviet Tupolev Tu-114 *Moss* early warning aircraft which was operative several years before the similar USAF Boeing E-3A AWACS. In fact, in 1971, it participated in the Indo-Pakistan conflict with Soviet crew, successfully guiding Indian fighters in the interception of hostile Pakistani aircraft.

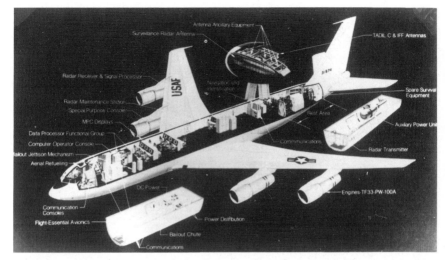

Cut-away of a USAF Boeing E-3A Sentry AWACS (Airborne Warning and Control System). This aircraft, like the KC-135, derives from the well-known Boeing 707 airliner.

USAF Boeing C-135 Stratolifter converted into a flying laboratory for high-energy laser tests. The dorsal fairing behind the cockpit and the tail boom contain antennae and equipment. An aircraft of this type blew up in the skies of Maryland on 6 May 1981 during secret experiments. (By courtesy of *Aviation Week & Space Technology*)

The General Dynamics EF-111A tactical ECM jamming version of the variable-geometry F-111 all-weather attack and FB-111A strategic bomber of the USAF is the most up-to-date aircraft equipped for electronic counter-measures tasks. The EF-111A can jam hostile radars even if they change frequencies since its jamming system automatically adjusts onto the new frequency. Its main task is to give ECM escort to nuclear attack aircraft during deep penetration missions into enemy territory and to assist in avoiding early warning radar.

Modern equipment for electronic warfare which can operate in an environment characterised by a great quantitiy of electromagnetic emissions. Highly automated, with computer-processing of all emission data, it can display the operating stage most suitable to the actual tactical situation. It permits one operator to control instantly all the fundamental functions of electronic warfare: interception, threat identification, jamming, etc. (By courtesy of Electronica SpA).

by comparing photographs and recordings made during each flight, it is possible to acquire important information regarding planned military installations, radar stations, missile ranges, submarine bases and so on.

The route Powers was to take was as follows: Adana–Peshawar (Pakistan)–Kabul (Afghanistan)–Sverdlovsk (Russia)–Bodö (Norway). Powers carried a ·22 calibre pistol and, hidden in a silver dollar in an inside pocket, a tiny syringe containing a lethal dose of the poison curare, for use in the event of a forced landing. The injection was, according to CIA instructions 'optional'; being a spy-pilot on a salary of $35,000 a year had its risks!

The first leg of the trip was simply a positioning flight from Incirlik to Peshawar, where he would stay for four days to rest and re-fuel before starting the long flight over Russian territory.

On 1 May, Powers got back into the cockpit of the U-2 to accomplish his 'mad flight' which would take him a distance of 3525 miles, flying at an altitude of 100,000 feet, over the Urals, the Russian cities of Stalingrad, Aralsk, Kirov, Archangel and Murmansk and the two important missile ranges at Tyuratam and Kapustin Yar, which had been recently discovered by US espionage services.

The U-2 took off about an hour behind schedule as the 'go-ahead' from President Eisenhower, normal procedure for all flights over Russia, was late arriving. While he was waiting, Powers had checked the equipment on board the plane several times and no doubt felt some apprehension when his eye fell on the button marked 'DESTRUCTION', to be pressed in an emergency to prevent certain items of top secret electronic equipment falling into Russian hands. According to the directions written beside the button, the explosive would destroy only the equipment itself but Powers knew that the explosive charges were attached to the inner walls of the pressurised part of the fuselage and, given the enormous pressure differential at very high altitude, the explosion would also mean the certain end of the aircraft itself.

After take-off, the U-2 began a rapid climb and, by the time it was over Kabul, the capital of Afghanistan, it had already reached an altitude of 65,000 feet; at this point, Powers switched on the equipment which could pick up and record all electromagnetic emissions present in the atmosphere, including military radio messages and radar signals, on all frequencies in current use. The equipment would automatically record the main parameters of every radar: frequency, duration of individual pulses, PRF and ARP (antenna rotation period). These parameters constitute the 'signature', a fingerprint, of a radar and, by analysing them, one can determine the type of the radar in question and the particular operative purpose for which it is employed. By taking two or more bearings (measurements of the direction of arrival of electromagnetic emissions made in such a way as to allow triangulation), one can determine the location of the radar and, consequently, where the weapons system it is used to control, is located. When the radar belongs to a potentially hostile country, this information is extremely useful for devising electronic measures and countermeasures for future use by one's own pilots who might have to penetrate enemy air space.

A primary use of such parameters is to store them in, or have them 'memorised' by an RWR which warns a pilot of the presence and direction of an enemy ground-based or airborne radar constituting a threat. Forewarning of an imminent missile or anti-aircraft artillery attack is, obviously, a vital factor in the success of a mission and the survival of the pilot himself who is thus enabled to make an immediate evasive maneouvre or effect the appropriate electronic counter-measure to deal with the threat. There are many different types of electronic countermeasures, and the choice of which to use depends on the contingencies of the particular situation. For example, the pilot can jam the enemy radar to neutralise its effectiveness or use electronic deception to send a missile off course by sending back a false electromagnetic signal to the radar guiding the missile. 'Chaff', which enjoyed such great success in World War Two, can be used to divert the missile from the real target by creating numerous false echoes in the vicinity of the aircraft. Chaff is now manufactured from various materials such as silver-plated nylon, lead-plated aluminium, aluminium–plated fibre-glass and other combinations. The strips are now launched automatically from the aircraft by various types of dispensers in patterns and manners for specific purposes.

Powers was not an expert in electronic warfare. However, as a U-2 pilot, he had undergone the usual CIA training and knew that, should a Russian MiG-21 flying about 30,000 feet beneath him, launch an air-to-air missile to try to shoot him down, he could rely upon the new, extremely sophisticated electronic apparatus installed on board the aircraft to confound the radar of the Russian missile. This apparatus was, in fact, one of the very first deception jammers (DJ). It had been specially designed, at the request of the CIA, by three leading American companies in the field of electronic warfare equipment and was, of course, 'top secret'.

Powers also knew that the Russians had been extremely annoyed by previous U-2 flights, although they had maintained a dignified self-restraint since they could not do anything about the matter. The Russian radars were no doubt lying in wait for him and would try to locate him as soon as the U-2 entered Soviet air space. However, Powers was comforted by the thought that his aircraft could fly at such a high altitude that no Soviet interceptor or missile could reach him.

US radar stations in Pakistan and Afghanistan were able to track Powers until he crossed over into Soviet territory and disappeared from their radarscopes. The only contact from then on was via CIA listening stations which intercepted Russian air defence radio com-

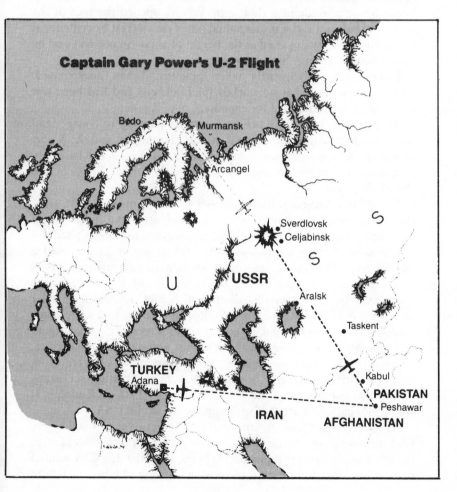

munications; no direct contact with Powers was possible as he had to maintain strict radio silence.

Shortly after Powers had left Afghanistan, a Soviet radar began to signal to other stations that it had detected an unidentified aircraft and, as the U-2 penetrated into the heart of the Soviet Union, news of the sighting was passed on from one station to another. Suddenly, CIA operators heard an excited voice saying over and over again in Russian, 'Target hit! Target hit!' At that very instant, Powers felt himself being flung forward in the cockpit and dazzling red light illuminated the fuselage of the U-2 from outside, as though there had been a violent explosion behind the aircraft. The aircraft went out of control and began losing altitude in a slow spin. Powers opened the roof of the cockpit to bale out and was catapulted out of the aircraft by centrifugal force. His parachute opened at the height of about 30,000 feet and he slowly descended into Soviet territory.

Powers was subsequently taken prisoner. When interrogated, Powers maintained that he worked for Lockheed and had been test flying an aircraft intended for espionage over Russia.

All this took place just before a summit meeting in Switzerland where the President of the United States, Eisenhower, and the Soviet Prime Minister, Krushchev, were to meet to discuss peace. Krushchev took advantage of the occasion to humiliate the United States before the whole world.

Powers was, naturally, tried in Moscow before a military tribunal of the Soviet Supreme Court where he was described as a typical unscrupulous young American who, in his greed for money, had not hesitated to commit a criminal action which might have caused a nuclear war. The public prosecutor introduced as evidence reels of magnetic tape recovered from the wreckage of the U-2. These contained the characteristic bleeps of Soviet air defence radar indicating the PRFs of the radars. Photographs of the U-2's electronic equipment were also shown, projected onto a screen in the courtroom.

The sentence was severe, although, as it was stated in the verdict, 'limited' to ten years' segregation, three years of which were to be spent in prison. Powers, however, was released after seventeen months, in exchange for the KGB master-spy Lt Col. Rudolph I. Abel, who had been arrested and imprisoned in the USA.

As soon as he set foot on American soil, Powers was grabbed by the CIA and subjected to incessant interrogation for a period of over twenty days; there were many points in the affair that the CIA wanted to clear up. They were most interested in learning whether the failure of the U-2's mission was due to the existence of new Russian surface-to-air missiles or to an act of treason committed by Powers and, above all, whether the Russians had acquired the means of preventing American bombers equipped with electronic devices from penetrating Soviet air space in the event of war.

During his trial in Moscow, Powers had claimed that the aircraft had been hit at an altitude of somewhere between 45,000 and 73,000 feet, instead of the 100,000 feet recommended altitude for that type of mission. Powers explained that the U-2's jet engine had stopped due to a fuel blockage and he had, consequently, lost a considerable amount of height while attempting to re-start it.

Other mysteries which needed to be clarified concerned Russian radar detection and the actual shooting down of the U-2. How had the Russians been able to locate the aircraft so quickly, given that it was made of radar-resistant material? And, if it had been hit by a missile, how had its photographic and electronic equipment remained intact? According to Krushchev, the U-2 had been fired on and hit by a single missile at about 65,000 feet. If this were true, the CIA asked themselves, why had the RWR and DJ failed to work?

Secret US sources of information inside Russia said that, just as Powers was about to over-fly the missile range at Sverdlovsk, the Russians had sent two MiG fighters to intercept the American aircraft and, immediately thereafter, had launched three surface-to-air missiles (SAMs). It would seem that two of the SAMs had hit the MiG fighters, shooting one of them down, but that the third missile had exploded near the tail of the U-2.

This information has never been verified but, if this were in fact what happened, a possible explanation could be that the U-2's ECM equipment had managed to send two of the SAMs off course but, saturated by the emissions from the first two missiles, had been unable to receive and successfully counter the electromagnetic signals coming from the third missile, which had continued on course towards the U-2. Nevertheless, considering that the U-2 had not received a direct hit from the missile, one wonders why Powers did not use his ejector seat, which, when it was fired, was probably designed to trigger the aircraft's self-destruct mechanism, to abandon the aircraft, instead of wasting precious time getting out under his own steam. And, if he had had so much time to spare, why had he not pressed the DESTRUCTION button and thus destroyed the top secret electronic equipment on board the aircraft?

After being interrogated by the CIA, Powers was also interviewed by various government commissions and even spoke to Congress but no satisfactory answers were given to the above questions.

The CIA even engaged one of its most attractive lady secret agents to try to make Powers talk, employing methods which are not exactly orthodox! But the outcome of this was simply that Powers divorced his beautiful wife, Barbara, and married the lady whom the CIA had hoped would bring him before a special tribunal.

A few years later, a link was noticed between the operations of U-2 aircraft and Lee Harvey Oswald, the man who assassinated the American President, John Fitzgerald Kennedy. It seems that, while in the US Marines Corps, Oswald had served as a radar controller in the air traffic control station at the American military base at Atsugi in Japan. In this capacity, he was not only able to observe U-2s taking off and landing but also had access to secret, high-level information regarding aerial espionage over the USSR and China. Since exchange of communications between a U-2 pilot and controllers at the air base before take-off was a routine procedure, Oswald was able to listen to requests for meteorological information for the specific routes and altitudes of the U-2 during its special missions. Oswald later defected to Russia where he stayed for a while until the KGB sent him back to America where he would be more useful to them.

The hypothesis that it was Oswald who furnished the Russians with information regarding U-2 routes and altitudes provides a fairly plausible answer to the questions cited above; that is, how had the Russian radars been able to detect and track the flight of an aircraft made of radar-resistant material and, secondly, how had the Russians been able to hit the U-2 with missiles whose range was inferior to the flying altitude of that aircraft?

This hypothesis also gives weight to the rumour that the U-2 was the victim of sabotage, and that, according to some reports, agents working for the KGB had placed a small radio-controlled or timed explosive device in the tail of the U-2 before take-off which had caused the aircraft to lose altitude when it exploded.

On 1 August 1977, Francis Gary Powers died tragically in a helicopter accident, at the age of forty-eight. The helicopter, which belonged to a Los Angeles television company, crashed in the middle of a forest-fire which the ex-U-2 pilot had been filming. His charred body was found and carried away in a sack, carrying away, at the same time, any possibility of clearing up the many mysteries of the flight which had caused such an unprecedented international outcry and which had remained an enigma even for the CIA.

The Case of the B-47 Stratojet

Immediately after the U-2 incident involving Powers, which took place on 1 May 1960, the American President, Eisenhower, suspended all U-2 flights over Soviet territory and ordered military authorities to devise other systems for collecting electronic information and photographing Soviet territory. The idea of using artifical satellites for such purposes no doubt arose at this junction; these could operate unmanned and would be out of the range of any weapons system then in existence.

However, such an ambitious project would take time to develop and, meanwhile, the CIA could not afford to be deprived of data concerning Russian radar systems, data held to be vital for the defence and even for the survival of the American nation. The head of the CIA, Allen Dulles, held that the real task of American intelligence was to carry out espionage activities behind the 'Iron Curtain' and that traditional means of doing this were no longer viable. He remarked that the KGB could buy for five cents, the price of *The New York Times*, information which the CIA could not buy even for $10,000! In the USA, all projects concerning rockets, missiles and so on have to be approved by Congress and are therefore publicly discussed. All airbases are marked on ordinary road maps and news of any atomic explosions made in the desert of Nevada is published in every newspaper. The Russians, on the other hand, keep quiet about such things. The smallest and most insignificant piece of information requested by the Defence Department can mean the death of the American agent asked to supply such information. The gist of Dulle's argument, which he continually restated, was that the Americans let the Russians know too much while the latter did not let them know anything!

To overcome their disadvantage, the Americans had to accelerate their progress in the field of technology, especially electronics. New listening posts were set up in countries bordering Russia which were allies, or on friendly terms with, the United States. Their extremely advanced radio receivers could intercept up to two million words a day, which were immediately retransmitted to Washington to be deciphered.

In this way, the Americans had access to interesting Russian communications. For example, in 1958 they managed to listen to a discussion between two Russian fighter pilots as they were attacking a Lockheed C-130 Hercules which was on a ferret mission. In April 1967, they were able to follow the dramatic adventure of the first

Russian Soyuz spaceship with the cosmonaut, Komarov, on board. As the space capsule was hurtling towards the Earth, Komarov, who was alone on board, realised with horror that the controls which should have operated the capsule's parachutes were not working. On the ground, his wife and the Soviet Prime Minister, Kosygin, tried to keep up his morale by telling him that he had been awarded the country's highest honour, but Komarov continued to scream 'I don't want to die! Do something!' until the space capsule finally disintegrated.

Although Russia was the main focus of interest, the Americans also intercepted and deciphered all military, diplomatic and commerical communications of other key countries, especially in periods of international crisis. This type of activity comes under the authority of the National Security Agency (NSA). At this point, the Americans were able to intercept radar emissions from potentially hostile countries in any part of the world. Once the main characteristics of the radar had been determined, NSA electronic experts would physically reconstruct the radar in question for detailed analysis and training.

New long-range radars were installed in many bases surrounding Russia. These had a coverage of about 1000 miles into Russian territory and were able to follow experimental missile launches at Tyuratam and even those which took place at the missile range of Krasniy-Yar, 750 miles from the Turkish border. The radars were able to track a missile until it landed in the desert of Kuezuel Kumm near the Russian border with Afghanistan. Other new stations for intercepting, recording and analysing the emissions of Soviet radars were established wherever friendly countries granted permission for them to be set up.

However, in spite of all these provisions, many Russian radar emissions could not be intercepted for the simple reason that, Soviet territory being so vast, the emissions from radars situated in central Russia or Siberia did not reach the borders. It was necessary, therefore, to send aircraft to locate or confirm the locations of new Russian radar stations in areas far away from American intercept stations. Thanks to their mobility and altitude, these aircraft enormously extended the reception coverage, even though, after the recent U-2 incident, they no longer flew directly over Soviet territory and U-2s were no longer employed in such missions.

On 1 July 1960, an ERB-47 a version of the six-engine strategic Boeing B-47 Stratojet (ER standing for electronic reconnaissance), took off from the British base of RAF Brize Norton on an ELINT mission which would take it along the extreme northern coasts of the

USSR. The ERB-47 had a ceiling of 43,000 feet, a range of 3200 miles and a maximum speed of 725 mph. It was to follow a triangular route, starting from a point 100 miles west of the island of Novaya Zemlya, then fly parallel to the coast of this island until it reached the extreme north-eastern tip from where it would begin its return journey via the Barents Sea. The last radio contact made with the aircraft was when it was 300 miles to the west of Novaya Zemlya, where the Russians tested their intercontinental ballistic missiles (ICBMs) during the summer months.

The ERB-47 was located by Russian air defence radar and fighters were promptly sent out to intercept it. Five hours after take-off, the six men on board the American bomber, which was flying at an altitude of about 32,000 feet, saw the first MiG fighter flying overhead. Shortly afterwards, another MiG approached from the right and opened fire. The ERB-47 returned fire with its tail guns but was no match for the Soviet fighters which had no difficulty in shooting it down.

As in the U-2 affair, the news was released by the Soviet Prime Minister, Krushchev, who once again accused the United States of violating Soviet air space. The Russians maintained that they had intercepted the aircraft 22 kms (13.5 miles) north of Cape Svyatoy on the Kola peninsula and that they had shot it down because it was heading for the major Russian port of Archangel. The Americans, on the other hand, held that the aircraft had been shot down 50 miles north of Cape Svyatoy.

A few hours after the American aircraft had been shot down, Soviet ships began to search for survivors in the Barents Sea. A trawler picked up two survivors, First Lieutenants John R McKone and Freeman B. Olmstead and the body of one of the pilots; no trace was found of the rest of the crew.

The two surviving officers were tried for espionage, convicted and imprisoned. They were later released, on 25 January 1961, following a personal intervention by the new US President, John F. Kennedy.

The Drama of the Spy-Ship *Pueblo*

In 1963, the Americans also began to conduct electronic espionage missions in Asia. Sixteen missions were carried out by US ships along the eastern coasts of Siberia, China and Korea but the only incidents worth mentioning involved the *Banner*, a spy-ship operating with the *Winnebago* in the Pacific. As she was carrying out her final mission, she was molested by Russian warships, one of which trained her guns on the *Banner* and hoisted signal flags which, in International Signal Code, meant 'Halt or I shall open fire'. One of the Russian torpedo-boats came up very close but nothing happened. On another occasion, Chinese trawlers surrounded the *Banner* and trained their guns on her. The captain of the *Banner* handled the situation brilliantly by steaming full speed ahead at the trawlers, putting an end to their harassment.

On 1 December 1967, USS *Pueblo* arrived in the Japanese port of Yokosuka, where American spy-ships were based. She was coming from a major refit in the USA during which radical modifications had been made to convert her from a supply ship into a ship for 'auxiliary general environment research', as was indicated by the letters AGER 2 painted on both sides of her bow. This was the official classification of the *Pueblo* and, to give credence to this, two civilian physicists and special equipment for oceanographic research had been placed on board the ship before it had left the USA. However, the true purpose of the *Pueblo* was (SIGINT), in other words the collection of data regarding electronic warfare. Eight antennae covered by radomes dominated the superstructures of the ship and, in a large ELINT operations room under the bridge, there were two large intercept receivers capable of picking up any electromagnetic emission coming from radio or radar even at great distances. All transmissions were automatically and very accurately recorded on special tape by the latest digital equipment. These tapes were later taken to CIA centres for analysis and evaluation.

The *Pueblo* displaced 900 tons, was 53.20 metres long and 9.75 metres wide and had a maximum speed of 9 knots. Her captain, 39-year-old Lloyd M. Bucher, was not a graduate of the prestigious Naval Academy of Annapolis; he had, in fact, been educated at a 'Boys' Town' boarding-school in Nebraska from whence he had joined the US Navy and, after graduating from the University of Nebraska, became a naval officer. The security officer was 21-year-old Timothy L. Harris, in charge of all intercept activity and related secret documents. The crew totalled eighty-one, and comprised six officers, twenty-nine ELINT operators, the two physicists mentioned above, and forty-four seamen.

At the end of December 1967, the *Pueblo* received her orders from the Commander-in-Chief US Navy in Japan to proceed on her first mission of electronic espionage; this entailed intercepting radio and radar emissions from North Korea and observing Soviet naval manoeuvres in the Tsushima Straits.

On 5 January, the ship departed from Yokosuka and, sailing past the island of Kyushu, arrived at the Japanese port of Sasebo on 9 January. Here, Bucher received detailed instructions regarding the mission and information about Russian ships which he might encounter.

On 11 January, before daybreak to avoid being seen, the *Pueblo* left the port of Sasebo and headed for the Tsushima Straits and the Sea of Japan where she was to carry out her mission. Her instructions were to record the radar emissions from North Korean coastal defence systems so that the United States could devise ways of neutralising these radars in the event of war. The captain's plan was first to collect electronic information concerning North Korean radar and then to observe Russian naval manoeuvres on his way back to Sasebo. He was authorised to approach 'not closer then 200 metres' to Russian warships for the purpose of obtaining close-up photographs. The operating zone was between the latitudes of 39 degrees and 42 degrees North. Orders were to maintan strict radio and radar silence; only in an emergency was the use of radio permitted. The reason for this silence was, of course, to avoid detection by Russian ships, or by patrol ships from other potentially hostile countries.

After several hours of navigation, the *Pueblo* headed due north towards the island of Ulung Do (marked 1 on the map on page 141) but ran into a violent storm en route and was forced to reduce speed and change course to avoid foundering. When the storm had passed, Captain Bucher headed towards his first objective, the waters outside the North Korean port of Chongjin (marked 2 on the map). He arrived there on 16 January and stayed in the area for two days, sailing against the wind at minimum speed, almost at a standstill, while electromagnetic emissions were intercepted and recorded. During the day, the *Pueblo* usually kept at a distance of 14–18 miles from the coast but, at night, she withdrew to 20–25 miles because of the difficulty of determining her exact position in the darkness. At regular intervals, the ship's engines were stopped so that the two scientists on board could carry out their oceanographic research, measuring the sea temperature and collecting water samples. The data provided by this kind of research is very important in anti-submarine warfare because it can be used to determine how temperature, salinity and other physical characteristics of the sea water in that particular area affect the performance of sonar, used for submarine detection.

After leaving the waters of Chongjin, the *Pueblo* headed south and, on 18 January, arrived near Songjin (3) where she remained for about two days without noting anything of particular importance.

The ship then sailed towards Mayang Do (marked 4 on the map) where she stayed until 21 January; that evening, just as the sun was going down, Bucher sighted a North Korean submarine-chaser sailing at a speed of about 25 knots. Thinking that the ship probably had not noticed the *Pueblo*, he decided not to signal the sighting to his Command in Japan since the radio communication might alert the North Koreans to the *Pueblo*'s presence. It would seem that Bucher gave immediate orders to leave the area and set course for the important North Korean port of Wonsan (marked 5 on the map). Weather conditions were very bad with continuous high winds and snow; nevertheless, the mission was, at this point, proceeding according to schedule. The *Pueblo* arrived off Wonsan on the morning of 22 January and, as usual, set about its task of intercepting and recording coastal radar emissions, always keeping clear of North Korean territorial waters, at least according to the ship's navigation officer, which began 12 miles from the coast.

At about 13.30, a seaman on watch signalled that two trawlers had left the port and were heading towards the ship. When they arrived at a distance of about 50 metres from the *Pueblo*, they began to sail around the foreign ship slowly. They were unarmed but the unwelcome visit was obviously the consequence of the previous day's encounter with the North Korean submarine-chaser.

Bucher ordered all hands to remain below deck so that the North Koreans would not see how many men were on board—certainly there were an unusual number for a ship that, to all intents and purposes, was supposed to be carrying cargo! At the same time, he ordered a signal to be sent to the US Navy radio station at Kamoseya in Japan informing them that the *Pueblo* had been discovered by the North Koreans.

For hours, the operators tried to transmit their message on the *Pueblo*'s WL-7 transceiver but, for some unknown reason, they were unable to do so. Meanwhile, followed by the two trawlers, the ship continued to navigate slowly at about 15 miles from the entrance to the port of Wonsan. At 9.00 on 23 January after about sixteen hours, *Pueblo*'s message to Kamoseya was finally transmitted!

Around midday, a North Korean navy SO-1 submarine-chaser arrived, sailing at full speed, with her guns already manned and trained-on the *Pueblo*. The North Koreans sailed once around the ship to get a close look at her and then asked her to communicate her nationality. In the meantime, four North Korean torpedo-boats had left the port of Wonsan and were approaching at full speed. When, in

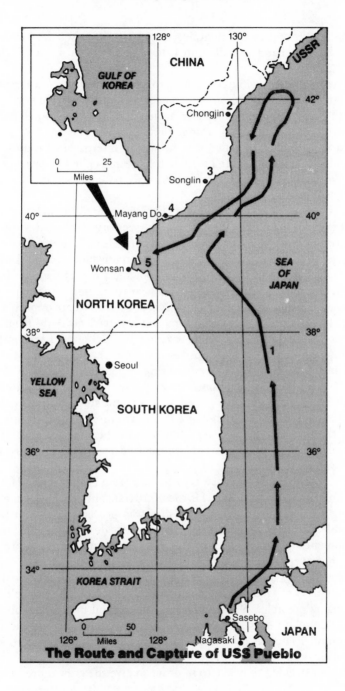

The Route and Capture of USS Pueblo

answer to the request for identification, the *Pueblo* raised the American flag, the sub-chaser signalled, in International Signal Code: 'Halt or we shall fire'.

The Americans, while continuing to steam slowly out to sea, replied by signalling that the *Pueblo* was an oceanographic ship. However, the North Korean submarine-chaser, which had now been joined by the torpedo-boats, ordered the *Pueblo* to follow her.

Captain Bucher signalled that he was in international waters and intended to sail out to sea. The North Koreans replied by firing at the *Pueblo*, wounding, although not seriously, two sailors and Captain Bucher himself.

It was now 14.20. At this point, Bucher ordered the helmsman to steer towards the port of Wonsan and the security officer to destroy the electronic equipment and secret documents. He consulted some of his officers, to see what could be done in the circumstances, and asked for the exact position of his ship at that moment; it was 15.6 miles off the island of Ung-Do which was situated near the entrance to the port of Wonsan. Bucher signalled that the North Koreans were opposing his right to sail in international waters but received no reply from the submarine-chaser, which was now sailing parallel to the *Pueblo*, while the four torpedo-boats were positioned on either side, one at the stern and one at the bow of the America ship. The officer in charge of the *Pueblo*'s weapons, which consisted of two 40 mm guns, reported to Bucher that they were encrusted with ice and still covered by their protective tarpaulins which made them extremely difficult to fire. In fact, the temperature was very low and Bucher realised that, if he scuttled the ship, the crew would not survive five minutes in those ice-cold waters.

The *Pueblo*'s only chance was to get help from US air or naval forces stationed in South East Asia. He therefore transmitted a radio message asking for help. Meanwhile, the ship was proceeding as slowly as possible in order to give the crew time to destroy the electronic equipment and secret documents; this operation was not going too well, however, as there were huge quantities of material to be burnt and insufficient devices to burn it with. Finally, an answer came from the US Navy in Japan saying: 'Your message received. Try to hold out as long as possible. We have ordered Command in South Korea to send Republic F-105 Thunderchief fighter-bombers. Good luck!' A few minutes later, as if by a joke of fate, two North Korean MiG-21s flew over the *Pueblo* and disappeared over the horizon.

Bucher decided to heave to in order to give the crew more time to destroy the secret material which, on no account, should be allowed to fall into enemy hands. The North Koreans on board the submarine-chaser reacted by firing the 57 mm machine-guns, wounding other members of the crew, while the torpedo-boats aimed their torpedo-launching tubes at the *Pueblo*. After a short while, one of the torpedo-boats moved alongside the *Pueblo* and about ten soldiers, armed with machine-guns and bayonets, boarded the ship. They were led by an officer who, gun in hand, started to give orders to the bewildered American sailors.

Thus did USS *Pueblo* surrender without a fight. Besides the dishonour, that surrender did untold harm to the United States because, together with the ship herself, the most technologically advanced electronic warfare equipment and the most top secret military documents then in existence fell into the hands of the communist North Koreans.

As soon as the news of the capture of the *Pueblo* reached Washington, the President of the United States, Lyndon B. Johnson, was woken up in the middle of the night and informed of what had happened in the Sea of Japan. As was his wont, Johnson replied, 'Thank you', and went back to sleep.

The Commander of the 5th Air Force was informed of the incident by telephone and, at 15.55, immediately ordered his Command in Okinawa to prepare all available aircraft to be sent to Wonsan; but, as these carried only nuclear weapons, they could not be sent, for obvious reasons.

The Commander-in-Chief of the Pacific Area ordered Fleet Command to send a destroyer to free the *Pueblo*, but it could not reach the area until 12.00 the following morning.

As soon as the Commander of the US Navy in Japan, Vice-Admiral Frank L. Johnson, Bucher's immediate superior, received the news, he rushed to headquarters in Tokyo and, on his own initiative, ordered the nuclear-powered aircraft-carrier USS *Enterprise*, then at a distance of 600 miles from Korea, to head for Wonsan. Armed aircraft could not be sent from Japan because of an agreement made with the Japanese government forbidding American military aircraft to fly combat missions from Japanese soil. Vice-Admiral Johnson considered that it would be useless to send rescue aircraft because it did not seem that the *Pueblo* was in danger of sinking.

On 23 January at Wonsan, the sun set at 17.41 and it was dark by 17.53, by which time it was already too dark for either aircraft or ships to come to the aid of the *Pueblo*.

No concrete help came from Washington, either. In spite of the pressure of public opinion, no military action was taken against North Korea. Only a formal protest was made to North Korea, while an appeal was made to the United Nations Security Council calling for the immediate release of the ship and crew, whose capture the US Government considered to be simply an act of piracy as the *Pueblo* had been in international waters.

So, for one reason or another, the *Pueblo* was abandoned to her fate! The members of her crew were imprisoned for almost a year. On 22 December 1968, they were released and allowed to return home, with the exception of one crew member who had died as a result of injuries received during the seizure of the ship.

Two days after the crew had been returned the Commander-in-Chief Pacific Fleet ordered a Court of Inquiry to be set up to investigate the circumstances relating to the capture of the *Pueblo*. A sub-committee, made up of high-ranking officers of the three US armed forces, was assigned the task of making a preliminary evaluation of the implications for national security resulting from the loss of the classified material which had been on board the ship.

The Loss of the EC-121

During the course of these hearings, the Pentagon suddenly announced on 14 April 1969, that, at midnight, the military forces of the Republic of North Korea had shot down a US Navy EC-121 which had been on a mission of electronic reconnaissance, about 50 miles off the North Korean coast. Given the similarity to the *Pueblo* incident, from a national security viewpoint, the sub-committee of the three armed forces decided to extend its inquiry to cover the loss of the EC-121. In fact, many parallels could be drawn between the two incidents and both revealed serious shortcomings in the chain of command.

The EC-121 was part of a reconnaissance squadron under the command of the Pacific Seventh Fleet and the Commander-in-Chief Pacific. However, the commander of the 5th Air Force was responsible for supplying aircraft to protect the EC-121 if necessary. However, when the spy-plane took off from Atsugi in Japan at 17.00 on 14 April 1969, it also departed from the operational control of the squadron command, but no other command took over operational control, even though various US Air Force, Navy and Army radar control centres followed its flight on their radar screens and tactical plots.

The first indication that the EC-121 was in danger came from the squadron command officer then on duty who reported that the command's radio station had intercepted a message from another American radio station warning that hostile aircraft were approaching the EC-121 in the skies over the Sea of Japan. The squadron commander then requested that the main American radio station in the Far East, at Fuchu, send copies of all radio messages transmitted by the EC-121 and, using all available sources of information, clarify why the mission had been interrupted. For over an hour and a half, the squadron commander called the radio station at Fuchu but no clarification of the matter was forthcoming. He therefore decided to transmit a lightning message, which would take precedence over all other transmissions, asking all relevant US command radio stations for news of the EC-121.

Immediately afterwards, squadron command received a message saying that the EC-121 might have been shot down by North Korean fighters over the Sea of Japan. At this point, the squadron commander asked the 5th Air Force to organise a rescue mission immediately and received a confirmation that a C-130 Hercules was being made ready for such a mission. The local time was 1.09, 15 April, and, probably because of the darkness, no trace of the EC-121 or its twelve crew members was found.

The Navy Court of Inquiry into the capture of the *Pueblo* was composed of five admirals and was presided over by Admiral Harold G. Bowen. The crew of the *Pueblo* and everybody who had been directly or indirectly involved in the mission were interrogated over a period of two months. All five admirals had fought in the Korean War, one of the toughest wars ever fought by the United States. They had been chosen for the Court of Inquiry for this very reason and, naturally, were not very lenient with Bucher. Bucher had surrendered his ship to the enemy without resistance and this, for them, was unforgivable; a Captain must never surrender his ship, whatever the circumstances. As a last resort, if he is really out-numbered, the ship should be scuttled The verdict was severe; the Court asked that Captain Bucher be brought before a Court Martial and tried on five charges: allowing his ship to be searched when he still had the means to resist; failure to take immediate offensive action when attacked by the North Koreans; obeying North Korean orders to follow them to the port of Wonsan; failure to ensure, before taking to sea, that his officers and crew had been prepared and drilled for the destruction of the secret documents and electronic equipment on board the ship; and failure to destroy these documents and the equipment through negligence, thus allowing them to fall into the North Koreans hands.

The Court also asked that Vice-Admiral Frank L. Johnson, Commander-in-Chief of US Naval Forces in Japan, be reprimanded for not having ensured that the *Pueblo* was adequately prepared and protected and, likewise, that Captain Everett B. Glanding, Security Chief of the Pacific Command, be reprimanded for not having confirmed that the efficiency of the *Pueblo*'s data-collection section was adequate to meet the severity of potential demands upon this vital section.

However, on the very same day that the Court issued these recommendations, the Secretary of the Navy sent out a communique stating that no action would be taken against the crew of the *Pueblo*, as they had suffered enough during their imprisonment, and that no judgement could be made either to absolve or condemn the officers and the captain since the premise upon which the activity of ships like the *Pueblo* was based—the freedom to sail in international waters—had been violated by the North Koreans' attack outside their own territorial waters.

The inquiry conducted by the sub-committee of the three armed forces to investigate electronic surveillance re-examined many aspects of the case dealt with by the Navy Court of Inquiry and, finally, drew up a report containing some very interesting revelations, conclusions and recommendations.

The *Pueblo* and the EC-121 Warning Star operations were part of a costly national defence plan to acquire military information about potentially hostile countries. According to experts in the science of modern warfare, national security is based on knowledge of the military capacities of potential enemies and, to acquire this knowledge, the best technical means must be employed to collect, analyse, evaluate and exploit information relating to these capacities. To this end, the United States had begun to conduct large-scale surveillance, both open and covert, using specially-equipped ships and aircraft to collect the necessary technical and operative information.

Both USS *Pueblo* and the EC-121 brought down by the North Koreans were being used for such purposes, and, as ship and aircraft respectively, had both had their advantages and limitations. However, generally speaking, both ships and aircraft had proved themselves to be extremely useful for this kind of activity, whether they operated separately or together.

In the early period of the Cold War, the US Navy used ordinary warships, cruisers, torpedo-boats and so forth, for collecting electronic warfare data. However, this practice was abandoned after a few years as there were several serious disadvantages: first, warships had to be taken away from their normal duties; secondly, the presence of a US warship in a sensitive area could be seen as a provocation by the country whose shore it was near which, therefore, limited the warship's capacity to undertake electronic espionage; thirdly, according to various treaties and international conventions, warships are subject to a number of restrictions which do not apply to other ships; finally, warships did not always have enough space for all the necessary electronic devices and for the technicians needed to operate them. Therefore, the authorities decided to use merchant ships for electronic espionage activity. In some cases, ships were designed and specially-built for such missions, in others, existing ships were suitably modified for their new role.

The first ship designed specially for electronic surveillance was ordered from a New York shipyard in July 1961. She was named USS *Oxford* and bore the letters AGER-1. She closely resembled the famous Liberty ships built during World War Two, especially her hull. Later, a further six ships of this class were ordered and named *Georgetown, Jamestown, Belmont, Liberty, Valdez* and *Muller*.

By 1965, these ships were judged to be too few to cover the national need for the collection of electronic information and so the US Government authorised a number of auxiliary ships, like the *Pueblo*, to be converted into spy-ships. Built during World War Two to fulfil US Army maritime transport requirements, they had been decommissioned in 1944 and placed in reserve. The first two ships to be refitted were the *Banner* and *Pueblo*, followed by *Palm Beach*. The US Navy was highly satisfied with this type of spy-ship and plans were approved for the deployment of fifteen such vessels in seas all over the world. Another factor which contributed to the US Navy's enthusiasm was the much lower cost of using such ships compared to other types.

The main advantage of using surface ships for electronic reconnaissance, according to the US Navy, was their ability to remain on station in one area for great lengths of time (a ship of the 'Pueblo' class had an endurance of 4000 nautical miles!) and are therefore bound to pick up new enemy radar signals sooner or later. Another great advantage was that such ships are protected by international conventions signed by all countries in the world, which state that a ship is part of the territory of the nation whose flag it flies and that it therefore cannot be attacked or captured. Finally, as mentioned above, the financial aspect was favourable.

In practice, however, the *Pueblo* did not possess any of those qualities which had initially made the US Navy so keen to use ships of this type; on the contrary, she was not 100 percent sea-worthy, and she was poorly armed, slow and unreliable with hopelessly inadequate devices for the destruction of secret documents and equipment. These factors were probably the real reason why no action was taken against the captain of the *Pueblo* and his crew after the inquiries. Moreover, the orders Bucher received were vague and incomplete and there was not even a semblance of organised assistance for the ship in an emergency.

The lesson to be learned from both the *Pueblo* and the EC-121 incidents is that, the more difficult the mission, the clearer and more unequivocal the chains of command need to be. This is absolutely vital because gaps in the chains of command at critical moments can lead to disastrous consequences.

Another lesson to be learned, particularly from the *Pueblo* incident, is that such vessels must have an appropriate defensive capability. They must be adequately armed in order to be able to defend themselves; they must be equipped with suitable early warning systems in order to sight a potential enemy before being sighted themselves; and they must be fast enough to get away from a danger area quickly, before they run into serious trouble.

As we have seen, ships like the *Pueblo* operated as part of an integrated programme of electronic surveillance and espionage drawn up by the US Navy in 1965. Those stationed in the Pacific were under the overall command of the Commander-in-Chief Pacific via the Commanding Officer Pacific Fleet who gave operating orders directly to periferal naval high commands. The *Pueblo*'s mission was part of the general plan to cover areas, like Japan, where there was a notable shortage of information regarding electronic warfare systems. Missions carried out in that area, therefore, came under the operative control of the naval command in Japan. But, at the moment of need, no command seemed able to take a decision.

There are still many unanswered questions concerning the capture of the *Pueblo* and the shooting down of the EC-121. One question of immediate interest regards the position of the *Pueblo* at the moment of capture; was she outside or inside North Korean territorial waters?

As all sailors know, it is not always possible to determine with absolute certainty the exact position or 'fix' of a ship, due to various factors such as wind, sea currents, the lack of conspicuous landmarks along the coast, and the reliability of the ship's navigation system. As a result, frequent disagreements arise regarding the position of a ship near the limits of territorial waters. In the case of ships like the *Pueblo*, the main reason for this uncertainty was the lack of precision of the navigational systems used. The *Pueblo* used the Loran (Long-range navigation) system which fixes a position by transmitting synchronised impulses from various radio stations situated at great distances from one another. With Loran, or, for that matter, any other similar radio-electric system, there can be a margin of error of several miles, especially near the coast, and, consequently, whether the *Pueblo* was inside or outside the territorial waters of the republic of North Korea is simply a matter for conjecture, and the truth will never be known.

Another important aspect of the *Pueblo* incident is the question of responsibility. After transmitting a radio message informing his command that the ship had been discovered by the North Koreans, Bucher had to wait almost twenty hours before he received a reply. This fact, coupled with Bucher's failure to take any initiative, is perhaps the main reason for the loss of the ship.

The case of the EC-121 was somewhat different. When a slow, unarmed and otherwise undefended aircraft is attacked, the crew do not have many options open to them. Consequently, active operational control of the aircraft by higher authority is of even more crucial importance. Those who planned the EC-121's mission should also have provided for its defence. Two main errors were committed: first, after what had happened to the *Pueblo*, an undefended aircraft should not have been sent to operate in an area where it was quite likely to be attacked and where it would be difficult to intervene in an emergency, given the precarious operational situation of the US Air Force which was already involved in the Vietnam war; secondly, the responsibility for the operational control of that admittedly difficult mission had been divided between too many different commands, with the result that, at the crucial moment, it was not clear who was responsible for the aircraft and, consequently, nobody did anything to defend or rescue it.

One thing that the two episodes have very much in common is of an extremely serious nature: the failure by those who had ordered the missions to take subsequent responsibility for their outcome.

Modern Espionage

Modern secret service agents are somewhat different from those characterised in spy stories. Nowadays, the pilot of a spy-plane can collect more information in a single mission than a hundred traditional spies, of the kind employed in World War one, could collect in the course of a year! Those famous tales of beautiful women hiding in their throbbing bosom precious, minute maps stolen from the room of some sentimental captain or womanising general now belong to the past. That is not to say that traditional forms of espionage are no longer useful. On the contrary! One famous case of classic espionage occurred in the post World War Two era when Russian agents managed to obtain British atomic secrets from the British scientist Klaus Fuchs, who has been called the the 'spy of the century'. However, perhaps this was an exceptional case as the British traitor was strongly influenced by communist ideology.

This type of espionage is called 'penetration' because it involves penetrating the operative centres of a potentially hostile country by placing an agent inside or close to these centres who can then steal important documents or listen to discussions regarding the country's defence. This form of espionage is extremely difficult to carry out because there are strict checks and security measures to prevent the placing of external agents. This difficulty has been largely overcome by using internal agents, people who already have a position in such organisations and who, for ideological reasons or for money, aspire to work for enemy intelligence services.

However, apart from very important documents and plans, most of the information desired can be obtained by intercepting and deciphering enemy communications and, above all, by conducting electronic and photographic reconnaissance missions. With the development of extremely sophisticated photographic and electronic equipment for long-range monitoring of nuclear tests and missile-launchings, espionage has become more of a technological and scientific discipline.

When, in the 1950s, the possibility of a sudden, catastrophic nuclear attack became a real threat, the only means of foreseeing such an event was by espionage. Foreseeing such an attack is extremely difficult since, unlike traditional warfare where aggressive acts are preceeded by the mobilisation of troops, tanks, ships, etc., preparations for a nuclear attack can be made in secret.

Consequently, secret services in the atomic age must have up-to-date information on the offensive capacities of other countries, above all those concerning nuclear weapons and their launching systems. This involves being informed of the deployment of guided-missile bases and of technological progress made in the enemy's missile guidance systems. Moreover, since the only way of discouraging the enemy from making a surprise attack is the threat of a massive reprisal, it is also important to know the enemy's defensive capacities in order to study and plan ways of penetrating his defences with a reasonable chance of success and the survival of one's own forces.

Although it is possible to hide missile-launchers in underground silos, and to camouflage missile-guidance radars and resort to many other cunning tricks to deceive the enemy regarding one's activities and intentions, nobody has yet found a way to keep hidden the electromagnetic emissions of a radar which is nearly always linked to a modern weapons system. Sooner or later, during installation of the weapons system, training of the radar operators or, above all, during experimental launchings, the radar must be switched on; it is then inevitably detected and located by electronic reconnaissance. It is as though the radar leaves its 'signature', or 'finger prints', in the atmosphere for anybody to pick up, as happens in classic cases of murder detection. Once the radar has been detected and located and its emissions have been analysed, appropriate ECMs can be devised to neutralise or reduce its performance at the opportune moment.

15

International Crises

The Cuban Missile Crisis

Since World War Two, there have been several major international crises, crises in which electronic warfare has played a crucial role. The most potentionally explosive crisis began in the late summer of 1962, when the US Navy spy-ship USS *Muller*, patrolling and listening in the Caribbean Sea, intercepted unusual radar signals coming from the nearby island of Cuba. The tapes on which these signals were recorded were immediately sent to Washington for analysis to try to identify this new type of radar. The Americans were dismayed to learn that it was a Soviet radar normally used to guide ballistic missiles with nuclear warheads.

To confirm this discovery, maritime patrol aircraft were sent to the Caribbean on ELINT missions while, in the course of a few days, extremely sensitive intercept receivers were installed along the south coast of Florida, their antennae pointing towards Cuba. All radio communications transmitted and received by the island of Cuba were also intercepted.

Shortly afterwards, on 14 October, a U-2 was sent on a reconnaissance mission over the island. The photographs, taken at an altitude of about 100,000 feet, were developed that very night and immediately examined by CIA experts. They were compared with other photographs taken over Cuba two years earlier—in January 1960—during a mission made by another U-2, using an infrared camera system. This mission had succeeded in photographing every inch of Cuba without arousing the slightest suspicion among Castro's defence units.

The photographs were subjected to close analysis and, the following evening, indications of an MRBM (Medium Range Ballistic Missile) base in the area of San Cristobal were discovered. Further reconnaissance flights over Cuba confirmed that the Russians had, in fact, already installed a number of such missiles and were in the course of setting up launch pads for longer-range missiles. The range of these missiles would be about 1000 miles which meant that the Cubans would be able to hit and destroy many American targets, including Washington, the Panama Canal and a number of Strategic Air Command (SAC) bases.

The US President, John F. Kennedy, was informed of this on the morning of 16 October and immediately conferred with his closest advisors, asking them to make an in-depth study of the dangers these installations represented for the USA and what action could be taken.

The working-group examined various proposals and these were discussed over the next five days while U-2 reconnaissance flights were stepped up. A few days later, the Soviet Foreign Minister assured President Kennedy that Russia had supplied Fidel Castro with 'defensive' weapons only.

On 27 October a U-2 on a reconnaissance flight over Cuba was shot down by a Russian SAM-2 *Guideline* missile and it's pilot, Major Rudolf Anderson, was killed. U-2s were immediately taken off such missions and replaced by Tactical Air Command McDonnell RF-101 Voodoos based in Florida. These could fly at twice the speed of sound at altitudes ranging from 50,000 feet down to tree-top level. They were equipped with electronically-controlled cameras and flare-ejectors for night photography.

After a brief but intense series of low altitude night and day flights by RF-101s, the Americans received confirmation, not only from the photographs taken but also from what the pilots had seen with their own eyes, that forty-two launch pads for MRBMs and twelve launch pads for IRBMs (Intermediate Range Ballistic Missiles) had been prepared, along with radar equipment to guide the missiles. It was also confirmed that Cuba had forty-two Ilyushin Il-28 jet bombers, 144 sites for SAM-2 missiles, forty-two MiG-21 fighters and several Russian-built missile-armed boats as well as about 20,000 Soviet military advisors and technicians. Ground intercept stations provided confirmation that the frequencies of the emissions previously intercepted by ships and aircraft were indeed those of radars normally associated with ballistic missiles. The presence of nuclear warheads was not confirmed; these, along with the missiles, were probably being transported by the numerous merchant ships which were then sailing from Russia towards the Caribbean.

On the basis of this irrefutable evidence, President Kennedy decided to take action, informing the American nation and its allies of what had happened and what was about to happen.

Of the few solutions open to him to favourably resolve the situation, Kennedy had decided on a naval blockade of the island. All ships carrying arms, whatever their nationality, would be prevented from reaching Cuba. To allow the Russians to save face, the action was called a 'quarantine'.

At that time, eighteen Soviet merchant ships. loaded with missiles and their accessories, were sailing towards Castro's island. A naval encounter between the Russians and the Americans in the Atlantic seemed inevitable and the whole world waited with bated breath for what might mean the outbreak of World War Three.

The approaching Soviet ships, escorted by submarines, were kept under constant surveillance by the Americans. When the first ships were stopped, inspected and asked to change course immediately, USSR authorities gave orders for the fleet to turn back.

Never since 1945 has the world been so close to a nuclear catastrophe as it was in that October of 1962. If the US Navy had not collected and evaluated the electronic information relating to the Soviet radars so promptly, thus allowing the Russians time to install more missiles in Cuba, the consequences for world peace would have been much more serious as, once installed, it would have been extremely difficult to effect their removal.

After this humiliating climb-down in the Caribbean, the Russians initiated a massive shipbuilding programme to strengthen their fleet. In charge of this programme was Admiral Sergei Gorshkov, Commander-in-Chief of the Soviet navy. Every year, new warships of all types, armed with surface-to-surface and surface-to-air missiles, were launched, swelling the Soviet navy which is now the second most powerful in the world. The wide range of electronic equipment which was gradually installed on the new Russian ships, revealed by the forest of antennae on the rigging, was very impressive both in quality and quantity. In order to avoid hindrance from enemy ECM, progressively higher frequencies and increasingly sophisticated technology were employed in the design of the new Soviet radars.

Parallel to the strengthening of Soviet naval forces and the spread of Moscow's strategic influence, efforts were also made to increase the merchant navy with its cargo and auxiliary ships, including oceanographic ships and large fishing trawlers. Many of these auxiliary ships, which were under the direct control of the navy and which, according to Admiral Gorshkov, were to be considered an integral part of Soviet naval power, were adapted for the purposes of electronic espionage. Referred to by NATO as AGI (Auxiliary, Intelligence Gatherer) vessels, these ships constitute the eyes and ears of Russian naval intelligence which, like the other Soviet armed forces, has to gather information regarding the deployment and operative procedure of the radar, radio, radio-navigational systems and so forth of potential enemies, beginning with NATO countries.

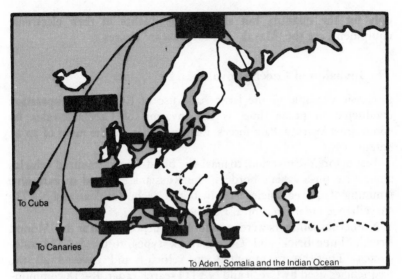

To Cuba

To Canaries

To Aden, Somalia and the Indian Ocean

Areas of Particular Interest For Soviet Spy-Ships and Usual Routes of Soviet Reconnaissance Aircraft

Since the NATO countries use many different types of radar and other radio-electronic systems and the area to be covered is fairly extensive, the number of AGI ships grew from four in 1962 to more than 160 in 1979. They are distributed among the various Soviet Fleets (the Pacific Sea Fleet, the North Sea Fleet, the Baltic Fleet, the Black Sea Fleet and the Mediterranean Fleet) and operate continuously wherever there are electromagnetic emissions to be intercepted. They are almost always present in an area where NATO air and naval exercises are being conducted, in the vicinity of missile ranges during missile test firings and along all coasts where NATO radar stations have been set up.

AGI ships vary greatly in their tonnage, range, endurance and so forth. For example, ships of the 'Primorye' class displace 5000 tons, have a large number of antennae on the masts and superstructure and two large rooms below deck which probably house electronic equipment for analysis of intercepted signals. They also possess a large number of receivers and transmitters for intercepting all communications made between NATO ships and their respective commands; they themselves can communicate with Soviet Fleet Command either directly or via satellite. Other classes of AGI, which are important not only for the quantity but also for the quality of their electronic equipment, are the 'Mayak' and the 'Okean' classes.

The Invasion of Czechoslovakia

A classic example of the large scale use of ECM in an operation conducted in peace time, is the invasion of Czechoslovakia by communist Warsaw Pact forces, mainly Soviet, on the night of 20/21 August 1968.

Just before the invasion, to mask the build-up of armoured vehicles along Czechoslovakia's borders, the Russians engaged in extensive jamming of all frequencies used by both Czechoslovakian and NATO surveillance radars in central Europe.

Numerous jammers were used for this purpose, such as the 'Mound Brick', 'Tube Brick' and 'Cheese Brick' types, to use NATO code-names. These were mounted on vehicles and covered all the frequencies used by Czech and NATO search radar. R-118 communication jammers were also used, mounted on trucks, to prevent or, at least, interfere with NATO and Czech communications.

Besides this jamming equipment, on the night of the invasion, the Russians also used large quantities of 'chaff' to completely blank out Czech and NATO radarscopes. Consequently, nobody was aware of the tanks advancing or the huge transport aircraft depositing men and arms at airports in Prague and other Czechoslovakian cities. The Russians managed to screen the presence of huge numbers of vehicles transporting the invading forces from all radars in the vicinity, thus maximising the element of surprise and their own safety in all phases of the operation. In short, this jamming operation completely paralysed any attempt at resistance, as the Czechs simply did not know what was going on.

The world was presented with a *fait accompli*. Western European countries and the United States could do nothing except follow a cautious policy of non-intervention and make a series of protests and accusations. The US Government also protested that the Soviet Union was trying to prevent, by jamming, 'Voice of America' transmissions from reaching the countries of the Eastern bloc.

However, the invasion of Czechoslovakia made American and allied military commands realise that their knowledge of Russian electronic warfare capabilities was severely lacking and they immediately stepped up electronic intelligence activity along their borders with Warsaw Pact countries.

The Build-Up of Soviet Electronic Warfare Capabilites

The Russians' near perfect use of ECM in the invasion of Czech-olsovakia was a real surprise for the Western powers and showed the importance the Russians attached to electronic warfare and the progress they had made in that field.

However, those who had read Marshal V. D. Sokolovkiy's book 'Soviet Military Strategy', published some years previously, should not have been so surprised. In this book, the ex-Vice-Minister of Defence of the USSR clearly specified the role of electronic warfare in his country's strategy. He defined the basic tasks of electronic warfare as preventing the enemy from effectively using electromagnetic emissions and protecting one's own emissions from enemy ECM. He wrote that ECM and ECCM were now in common use and their application was of the very greatest consequence, and that developments in the field of electronics were now of equal importance with developments in the fields of missiles and nuclear weapons, which themselves could be of little use without electronic equipment.

The organisation of electronic warfare in the USSR is very complex and is the responsibility of two large departments: the KGB and the GRU.

The KGB (*Komitet Gosurdarstarvenoi Bezopasnot*—Committee for State Security), is the senior of the two insofar as it comes under the direct supervision of the government. It collects every type of information pertaining to National Security by all possible means, from common spies to artificial satellites, from field intercept stations to stations installed in embassies and consulates abroad. It comprises four main Directorates, seven autonomous departments and six special sections. The KGB has a vast number of employees and huge material resources.

The GRU (*Gosurdarstarvenoi Razvedyvatelnaya*—State Military Information Agency), on the other hand, comes under the supervision of defence chiefs of staff and operates almost exclusively in the military sector. Like Western military information services, it deals with the gathering of operational and technical information regarding weapons systems, operational procedures and 'electronic orders of battle'[1] of potentially hostile countries.

In their electronic warfare operations, the Russians make wide use of airborne radar for the obvious reason that it has a greater range. Many different types of aircraft of varying sizes, have been modified for the purposes of electronic warfare. At first, civil aircraft were used, such as the Ilyushin Il-14 *Crate*, a twin-engined aircraft which entered service in 1954 in the passenger transport role, and the turbo-prop An-12 Cub, which entered commercial service in 1959 and was used for a while in Egypt to obtain information about Israeli electronic systems.

Certain types of fighter-bomber aircraft were also modified for the purposes of electronic warfare. One of the first of these was the MiG-21 *Fishbed-H* which carried its electronic equipment in a pod attached to the underside of the fuselage.

Another fighter-bomber used in electronic warfare has been the MiG-25, NATO code-named *Foxbat*. The high performance of this aircraft was an unpleasant surprise for the United States and countries of Western Europe when it began its development trials in the mid-1960s. At altitude, it could travel at Mach 3.2 (over three times the speed of sound) for short periods, although it was subsonic at sea level, had a ceiling of about 80,000 feet (24,400 m) and seemed altogether superior to its western counterparts. The photographic/electronic reconnaissance version first appeared in 1971 and was code-named *Foxbat-B* by NATO. Its sophisticated camera, IR-Linescan and side-looking radar (SLAR) and other electronic equipment aroused great curiosity and interest in Western intelligence services.

MiG-25s made regular flights over China and the Middle East. During the Yom Kippur War of 1973, the Israelis tried several times to intercept them using F-4 Phantom interceptors, armed with AIM-7 Sparrow medium range air-to-air missiles, but they could not get near enough to the Soviet aircraft to shoot them down. Moreover, attempts made by Western intelligence services to acquire information about the aircraft's capabilities were unsuccessful because the airbases where MiG-25s were stationed were subject to the strictest security. The aircraft was 'taboo' even among the Russians themselves and official documents referred to it as 'Product No. 84'.

Nevertheless, it was a Russian, a pilot named Viktor Belenko, who finally satisfied the desire of Western intelligence services to know more about this aircraft. On the morning of 6 September 1976, Belenko landed at the airport of Hakodate in Japan in a MiG-25 which had taken off from its base at Sakazovka, 190 kms north of Vladivostok in Siberia. Many other Soviet pilots had defected to the west in their aircraft but the one Belenko arrived in was something special! Electronic warfare and avionics experts were hastily sent from the United States to Hakodate to examine the aircraft's equipment. They dismantled the radar (a type, code-named *Jay Bird* by NATO, which operated on a frequency between 12,880 and 13,200 MHz), the

RWR (NATO code-named *Sirena* III), the ECM devices and the dielectric panels installed in the aircraft's nose to absorb radar emissions; all these were then subjected to close inspection and anlaysis.

Towards the end of 1976, information about the characteristics of the MiG-25 *Foxbat*'s electronic equipment was made known to all NATO Commands and the Soviet aircraft was no longer the terror of Western pilots.

However, a comparative evaluation of the *Foxbat* and contemporary American and European aircraft showed that Russian technology in this field, both in general construction and in its electronic equipment, was much less advanced than that of the Western world.

The two types of Soviet aircraft most commonly used for electronic reconnaissance have been the famous heavy bombers, the Myasischev Mya-4 *Bison* and the Tupolev Tu-16 *Badger*. The latter, in the *Badger-D*, *F* and *H* versions, is still in use today for collecting information. It carries an impressive load of eletronic and electro-optical equipment as can be seen from the many antennae (about a dozen) which, under their covers, protrude from all over the fuselage. The most recent version, the *Badger-H*, besides carrying a *Sirena III* RWR and passive receiving equipment (ESM) is also well-equipped with jammers for carrying out active ECM and can thus be used to provide EW support for raiding bombers.

Another important protagonist in the electronic war between Russia and NATO is the Tupolev Tu-95 *Bear-D*, an electronic and maritime reconnaissance version of the Tu-95 heavy bomber, which, with its four turbo-prop engines, has an endurance of 7800 miles without refuelling. It has often been seen flying over 'hot' zones during international crises to keep the situation under observation. The US Government has often complained of their presence in the Caribbean. Obviously in the region to intercept the electromagnetic emissions of the radars of new US naval vessels, they sometimes jam the radars of nearby US Air Defence Commands.

16

Vietnam: The Boom of Electronic Warfare

On 24 July 1965, during a raid over North Vietnam, an American McDonnell-Douglas F-4 Phantom was shot down by a Russian-built SAM-2 surface-to-air missile. It was neither the first time that a US aircraft had been shot down in Vietnam, nor the first time that a US aircraft had been shot down by a missile; five years previously, a U-2, piloted by Francis Gary Powers, had been shot down over Russian territory by a SAM. However, the loss of the F-4 was of great importance in another way because it marked the first appearance of Soviet missiles on the battlefields of Southeast Asia. Along with the missiles, the Soviets had also sent expert advisors to help the North Vietnamese.

The shooting down of the F-4 exposed the deadly threat constituted by Russian-built SAMs to the US Air Force, which had until then enjoyed air supremacy over North Vietnam. Vietnamese air defence assets had, so far, been limited to Russian-built MiG-17 and MiG-21 fighters and radar-controlled anti-aircraft guns. Now that ground-to-air missiles had also been deployed, North Vietnamese air defences were considerably strengthened.

US Air and Naval aviation losses had so far been acceptable, but now the situation changed dramatically. US aircraft found themselves without an adequate defence and losses began to increase daily. It was imperative to find an effective way of dealing with the new weapon.

In the USA, top level meetings were immediately held to study the problem. It was unanimously acknowledged that the only way of dealing with the new threat was to develop airborne electronic warfare systems to neutralise the radars used to guide the ground-to-air missiles. The task of developing such systems, assigned to leading US companies specialising in that particular field, was given top priority as a result of the alarming increase in aircraft losses over North Vietnam.

At the same time, great efforts were devoted to gathering technical and operational information about the SAM-2 missile system on the basis of which suitable antidotes could be devised.

The basic components of the Soviet SAM-2 system were the missile itself (NATO code-named *Guideline*), and the missile's tracking radar '*Fansong*'. Since its appearance in 1958, various modifications had been made to the system, particularly to the *Fansong* radar. In 1965, one SAM-2 system consisted of six missile launchers and one radar capable of guiding three missiles simultaneously. The whole system was transported on towed trailers and could be set up in about six hours.

The missile had a range of about 15 miles, a speed of Mach 3.5, and an explosive warhead weighing about 80 kgs. It had 'command'-guidance, a system in which the information needed to guide the missile is fed to it by an external source, in the case a radar.

In the SAM-2 system, the information was provided by the *Fansong* radar which locked onto the target and tracked it on frequencies between 2940 and 3060 MHz, transmitting via UHF radio the orders necessary to guide the missile onto target.

The *Fansong* radar also had TWS (Track-While-Scan) capability, utilising two radar beams, positioned at an angle to each other in the shape of a fan, which moved up and down like a bird flapping its wings. These beams were radiated by two antennae at right angles to each other which swept the sky from ground level to very high altitude and from right to left in an arc of 10 degrees each. This system permitted simultaneous coverage of an area of sky 3 to 4 kms wide and 3 kms deep around the target. The system also had a flat antenna which transmitted command signals for guiding the missile.

While they were waiting for US industry to develop appropriate ECM systems, the only chance of survival for the pilots of fighter-bombers operating in Vietnam was to try to evade *Guideline* missiles launched in their direction by violent maneouvres!

The gathering of electronic warfare (SIGINT) information had revealed several shortcomings in the SAM-2 system which could be exploited. For example, it took the *Guidline* missile a full six seconds, after launching, to be picked up by the tracking radar which would guide it onto target. Another limitation of the system was the missile's poor reception of the command signals transmitted from the ground and its slowness in executing the orders contained in the signals.

Exploiting the weak points of the system, the Americans came up with an evasive manoeuvre which immediately produced good results. It consisted in nose-diving in the direction of the SAM-2 battery as soon as the pilot saw a missile, or missiles, being launched. After its initial near vertical launch, the missile would veer downwards to get on course towards its target. At this point, the American pilot would suddenly pull up as hard as possible into a steep climb, flying inside the trajectory of the missile. Since the missile was incapable of making the violent manoeuvre necessary to 'capture' its target, the US aircraft usually managed to escape. However, this evasive tactic did not always work as clouds sometimes blocked the pilot's view of the missile.

By the end of 1965, the Americans, ever more deeply involved in the Vietnam war, had lost about 160 aircraft, most of them shot down by SAM-2 missiles.

Ground warfare in Vietnam was also difficult for the Americans because it was conducted according to the unorthodox guerilla principles clearly expounded by the leader of the People's Republic of China, Mao Tse Tung, in the following four rules, formulated many years previously:

When the enemy advances, we shall withdraw
When the enemy stops, we shall torment him
When the enemy avoids battle, we shall attack him
When the enemy withdraws, we shall pursue him

Vietnam was ideal for this new type of warfare and the Americans, who, for various reasons, could not use their nuclear weapons, found themselves in serious difficulties. There were no divisions or regiments of soldiers to confront in open battle but an ever-present, invisible army of men who could hide as they wished among the civilian population in houses, fields and, for the most part, in the endless jungle.

Not finding military or industrial targets worth hitting, the Americans directed their air attacks against enemy supply lines—most particularly the famous Ho Chi Minh trail, a secret route through the jungle and mountains of eastern Laos used for taking supplies from North to South Vietnam.

The Vietcong had dug, especially in areas around towns, numerous underground tunnels forming completely organised shelters. Many of the tunnels had concrete structures, first aid centres, warehouses, command centres, and so on; they even had electric light and a water supply and were provided with air by natural animal holes made by moles and rabbits. The entrances and exits to these tunnels were well camouflaged and, after a guerilla attack, the Vietcong would disappear underground like rabbits.

The tunnels had observation posts manned by sentries to keep an eye on what was happening above ground. When there was a low-level air raid, the Vietcong would prepare themselves and as soon as one or more aircraft had completed a pass and started to turn back, they would surface and fire at the disappearing aircraft, often managing to shoot them down.

In the early stages of the war, tear gas, and, possibly, other, more harmful types of gas, was used to drive the Vietcong out of their dens but, when news of this reached America and other parts of the world, there was a wave of protest which caused the Americans to curtail the practice.

The Americans then turned to other methods of detecting the presence of Vietcong. One of these methods employed common insects which can 'sense' the presence of human blood to which they are attracted. An electronic device was used to pick up and retransmit the 'emissions' which these insects made on sensing the presence of a man; these signals were then amplified so that the operators could hear them in their earphones. Another device used for the purpose of detecting the presence of the Vietcong picked up heartbeats and other physiological sounds made by the internal organs of the human body; these were transmitted directly to airborne aircraft.

The Americans also developed an electronic device capable of detecting the seismic vibrations produced in the ground by moving vehicles or troops. The devices were dropped from aircraft and their antennae, which were about a metre long, were camouflaged to merge in with the surrounding vegetation.

Another type of seismic detector, called an 'anti-intrusion' device, was used by foot patrols and small ground units to give them warning of a prepared ambush. The device consisted of small seismometers and a receiver which picked up their emissions. The troops planted and recovered the device themselves. If there were any 'intruders' in the area, the smallest vibrations caused by their footsteps would be detected, thus warning the troops of the presence of the enemy.

In order to reveal the presence of Vietcong guerillas at night or in thick jungle, the Americans used all the technical and scientific means they had at their disposal. One instrument used was able to detect the radio-electrical pulses emitted by the spark-plugs of engines, thus detecting the presence of motorised convoys, while magnetic detectors, by detecting variations in the magnetic field caused by the presence of metallic masses, were able to warn of the presence of arms or vehicles.

Another ingenious method involved making a chemical analysis of the air which, by detecting changes in the proportions of its components exploited the fact that human beings take in oxygen and expel carbon dioxide and nitrogen. If there were large numbers of Vietcong present, the air would contain proportionally slightly less oxygen and more of the other two elements.

Perhaps the strangest device for detecting Vietcong guerillas was one that emitted electromagnetic waves when touched or moved by somebody walking past it. The emissions were picked up by a small opportunely positioned transponder which amplified and retransmitted the sound-waves. The amplified signals were picked up by an electronic computer which processed the data which was then passed to commands.

For reasons relating to the frequencies used by these devices, it was nearly always necessary to set up a relay station to retransmit the signals to the processing centres. Specially-equipped aircraft were generally used as relay stations. The first type used for this purpose was the Lockheed EC 121R Super Constellation, a huge transport aircraft which could hold all the equipment necessary for processing the data transmitted by the small spy-devices scattered all over the jungle. The aircraft, flying at very high altitudes, covered vast areas and, on the basis of data received, could direct tactical fighter strikes on targets.

However, the use of Super Constellations proved to be too costly and, being unarmed, they were easy prey for enemy MiGs. They were therefore replaced by modified, suitably equipped single-engined utility aircraft. As the war progressed unmanned 'mini-planes' began to be used; these RPVs (Remotely Piloted Vehicles) also carried out reconnaissance and ELINT missions.

Tactical action taken by US strike aircraft against Vietcong infiltration through the jungle was often preceded by missions, using specialised aircraft, in which special chemical defoliants were sprayed on the surrounding vegetation. While facilitating the task of the strike pilots, this action caused almost irreparable damage to plant and animal life.

Processing of data acquired by the use of various types of detectors played a useful role both in tactical and strategic warfare in Vietnam. The following episode, described in a US command report, gives an idea of the tactical use of such data-processing.

In an area just south of the border between North and South Vietnam, a detector system had revealed the presence of an enemy unit which had infiltrated the area and was heading for a hilly area named Hill 881 by the military. Information concerning the movement of the unit, together with knowledge of guerilla fighting methods, enabled the US commander in that area to determine exactly the itineraries of the approaching enemy unit. Consequently, just as the enemy unit was lining up to attack, they were hit by an overwhelming artillery barrage from American artillery pieces hastily moved in to opportune positions.

However, the most important use of these data processing centres was in the field of strategic warfare, mainly in the interdiction effort to halt the flow of men and supplies along the Ho Chi Minh trail, the vital artery of the Vietcong. Contrary to what the word 'trail' suggests, this was not one, but a series of trails which crossed the Laos-Vietnam border over an area of about 100 kms. There were two main trails, each roughly 500 kms long, running from north to south and linked by a series of side roads running at right angles to them which afforded considerable flexibility in their utilisation.

To avoid large concentrations of vehicles, which would attract the attention of US aircraft which had already destroyed many bridges, road junctions and supply depots, the North Vietnamese had sub-divided the trail into sections, each one of which was the responsibility of a local command. Each command had its own vehicles and drivers who knew every inch of their stretch of road and could therefore get off the main route quickly at the slightest hint of danger. However, the disadvantage of this system ws that the vehicles had to be unloaded at the end of every section and the goods had to be kept well hidden until they were re-loaded onto the new vehicles of the next local command. These operations were, of course, always carried out at night. Infrastructures had been built, including rest areas and loading areas, and each transit station was protected by SAMs and AAA.

Early attempts to sever this umbilicial cord to South Vietnam by means of 'area' bombardments along the trail were not very effective. The Americans, therefore, adopted the tactic of 'selective' bombing which meant that the target had first to be clearly located and identified. It was in this phase that US forces made extensive use of the various detecting devices described earlier. After detecting the presence of a column of vehicles moving along the trail and determining its position, the column would be kept under observation and the time it would take to reach the point chosen for attack would be calculated. A number of control points were set up along the trail with various types of detectors: electronic, magnetic, infrared and so on. Wherever possible, the detectors were planted manually by experts; otherwise, they were dropped from aircraft, attached to metal spikes which would become embedded in the ground on landing.

The information gathered by this system—the 'Igloo White' system—was sent to a surveillance centre where it was processed by specially programmed computers and correlated with data acquired from other sources, such as deserters and spies. In this way, it was possible to plot the progress of the convoy and estimate its approximate speed, and to make an assessment of the number and the types of vehicles comprising the column. If, at a certain point, the calculated speed fell too far below average, it could be deduced that a transfer and rest station had been set up at that point. To determine the exact position of the station, further detectors could be launched, usually acoustic types, or infrared (heat-detecting) types which could detect the infrared energy emitted by vehicles or human bodies (see also Chapter 18).

Once the position of the stationary convoy had been established, the surveillance centre transmitted the relevant data either directly or via a relay aircraft to the aircraft which were to make an attack. This data was fed into the navigation computers by the pilots to calculate the course they should fly to reach and attack the convoy.

In 1966, the Vietnam war escalated and targets other than those along the Ho Chi Minh trail were bombed by the Americans. These new targets were often further north, sometimes as close as 10 kms from Hanoi, the capital of North Vietnam. As the distances that US strike aircraft had to fly to reach their targets increased so the bomb-load they were able to carry decreased and, consequently, B-52 strategic bombers, with their huge bomb loads, began to be employed. These had very sophisticated electronic equipment and flew so high that the Vietcong could not hear them. Consequently, the bombardments took the Vietcong by surprise and caused great destruction to their communication lines.

The year 1966 also saw the start of the great dogfights between MiGs and F-4 Phantoms over North Vietnam, such as the one that took place on 23 April between sixteen MiGs (fourteen Mig-17s and two MiG-21s) and fourteen Phantoms. The F-4s armed with passive radar homing AIM-7 Sparrow and IR-homing AIM-9 Sidewinder missiles and podded 20 mm cannons, were faster and, more importantly, had better manoeuvrability than the MiG-21—let alone the completely oudated MiG-17. The most feared enemy of the F-4 Phantom was neither of these. The real danger was from the SAM-2.

By the beginning of 1966, the US electronics industry had produced airborne electronic warfare equipment capable of intercepting pulse emissions from the *Fansong* radars and providing timely warning to the pilot that his aircraft had been locked-on and, within four seconds, could be hit by a *Guideline* missile. This RWR equipment was based on the 'crystal video' technique of signal detection, tuned into the

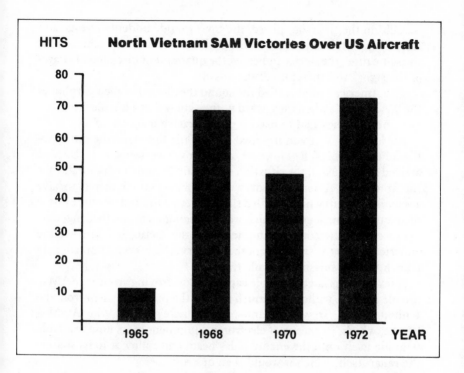

frequency band used by the enemy radars. As soon as the RWR picked up SAM-2 radar emissions, the pilot received an immediate warning.

RWR was first installed on modifed B-66 bombers. During air raids, an RWR-equipped B-66 would precede the formation, alerting it when SAM radar emissions were intercepted so that they could make appropriate evasive manoeuvres. Further progress made in the field of electronics led to the development of smaller RWRs which could also be installed on the strike aircraft themselves, although installation was still handicapped by the minimal space available inside such aircraft.

When the RWR's antennae picked up electromagnetic emissions from a radar, the system's receiver immediately passed them to a computer which compared their main characteristics with the stored parameters of radars acquired by electronic espionage. If the characteristics matched those of the SAM-2 radar, the receiver would immediately tune into those emissions, a red light would come on in the cockpit and the pilot would hear a signal tone produced by the radar pulses in his earphones; when the *Fansong* radar passed from the 'search' to the 'lock-on' phase, the tone would suddenly change and intensify due to the increase in the PRF as 'lock-on' was achieved. At the same time, the device indicated the quadrant of direction of arrival of the signal and therefore of the missile.

The American pilots called the sound they heard in their earphones the 'SAM song'; when they heard it, they knew that a missile was on its way and that they had to make a quick evasive manoeuvre.

Air losses in Vietnam dropped for a while following the introduction of this new RWR. However, as improved versions of the SAM-2 arrived from Russia, the situation once again became unfavourable for the Americans. American pilots were no longer content merely to have a warning system which enabled them to resort to a last minute evasive manoeuvre, a manoeuvre which was becoming increasingly dangerous because, the Vietcong having analysed the tactic, it often led the Americans into the sights of another battery which would immediately launch its own missiles (missile trap).

A few months later, a missile capable of destroying an entire SAM-2 missile battery, without too much risk for the pilots, was sent from the United States. It was an anti-radar missile (ARM), the AGM-4 Shrike, which, by means of electronic equipment in its nose, was able to guide itself onto the enemy radar beam and follow it to its source, the radar itself, which it would then destroy.

The new US tactic for destroying SAM-2 missile batteries involed sending between two and four two-seat aircraft, usually F-105 Thunderchiefs or F-4 Phantoms, to launch Shrikes. Each aircraft carried an EW operator whose task was to detect and locate SAM-2 batteries using an RWR to establish the direction of arrival of the radar beam and to guide the pilot towards it until it was time to launch the Shrike. At this point, there was no hope left for the SAM-2 battery as its own radar emissions would inexorably attract the very missile which would destroy it.

These mission were called 'Wild Weasel' and had a considerable measure of success. As a result of the effectiveness of the Shrike ARM and the RWR system, the number of US aircaft lost in proportion to the number of Vietcong missiles launched showed a significant reduction. Only forty U.S. aircraft were shot down by SAM-2 missiles in 1966, versus a great many SAM-2 batteries destroyed, in spite of the fact that the SAM-2 batteries had been considerably strengthened. According to American statistics, in 1965 one US aircraft was shot down for every ten missiles launched whereas, at the end of 1966, it was one in 70 (see the histogram on page 167).

In 1967 and 1968, more compact devices were installed on US fighter-bombers for maximum protection. The main problem with such installations was the lack of space inside the aircraft. This was solved by installing them externally in metal containers, or pods, hooked under the wings or fuselage.

The first type of device to be installed in this way was a simple noise-emitting electronic jammer. Subsequently, more sophisticated jammers were installed; these were able to evaluate the threats posed by the various radars, establish priorities and jam accordingly. Automatic chaff-launching devices, with similar evaluative capacities, were also installed in pods.

The huge B-52 bombers also benefitted indirectly from the new ECMs as they were always escorted by 'Wild Weasel' units in their air raids over North Vietnam.

However, American supremacy did not extend to the situation on the ground. The Vietcong were gaining more and more control of South Vietnamese territory and attempts made to weaken support for the Vietcong by means of psychological warfare had proved useless. Such 'psy-war' attempts involved air dropping thousands of propaganda leaflets; going into the villages to talk to the people; using special devices which emitted terrifying noises in the hope that the Vietcong would think it was the evil spirits of the forest; fitting loudspeakers to small aircraft which flew low over the villages, urging the rebels to surrender; and using Vietcong deserters to try to persuade their comrades to desert and go back to their families.

However, these forms of psychological warfare had little success. The Vietcong were far more successful in their regular mortar attacks on American airfields in which numerous aircraft and helicopters were destroyed.

In 1967, in an attempt to take the pressure off South Vietnam, where the Vietcong were becoming more and more enterprising, the Americans decided to bomb Hanoi, the capital of North Vietnam, and Haiphong, the most important port in North Vietnam. Airbases where MiG-17s and MiG-21s were based also became targets for air raids but the North Vietnamese government soon transferred these aircraft to airports on Chinese territory from where they could operate freely without fear of attack. Shortly afterwards, the North Vietnamese strengthened their air defence system with arms and equipment provided by Russia and China; this system included SAM-2 missiles, anti-aircraft guns and fighter aircraft, all coordinated by one central command.

With the success of US 'Wild Weasel' attacks on SAM-2 missile systems, the North Vietnamese increased the number of radar-controlled anti-aircraft guns until they had about 10,000 altogether. Most US aircraft lost in 1967 were, in fact, shot down by anti-aircraft gunfire while flying low to bomb bridges, roads, military bases and factories.

The SAM-2 missile sites were concentrated mainly around Hanoi, to defend the North Vietnamese capital from high-flying B-52 bombers: there were by then, only about thirty SAM-2 batteries still in operation. On the whole, the North Vietnamese air defence system was comprehensive and well-organised, covering practically all their territory and operating with unprecedented efficiency.

The bitter struggle between anti-aircraft defences and US aircraft was interrupted every now and again when the Americans suspended their air raids in the hope of being able to resolve the conflict by political means, allowing them to pull out of this messy war which was becoming more and more unpopular in the United States.

However, the war went on, and the US Navy F-4 Phantoms and A-4 Skyhawks were sent into action over North Vietnam, operating from the US Navy carriers *Kitty Hawk* and *Ticonderoga*. These aircraft were normally escorted by Douglas EA-1E (AD-5Q) 'Queer Spads' (Skyraiders) and Douglas EA-3A (A3D-1Q) Skywarrior ECM aircraft equipped with the latest electronic warfare devices, which greatly facilitated penetration of enemy air defences.

As the destruction caused by the Vietcong's nightly mortar attacks on US airfields in South Vietnam increased, operations carried out by these Navy aircraft were stepped up likewise. In the Gulf of Tonkin, there were twenty-five ships of the US Seventh Fleet, four of them aircraft carriers carrying a total of 600 of the latest strike/attack aircraft.

However, it soon became apparent that EW-equipped escort aircraft, which were usually slower than the strike aircraft they were escorting, were insufficient to reduce losses further. To achieve this, it would be necessary to improve the electronic equipment on board the strike aircraft themselves.

Further improvements were therefore made to RWRs as the pilots of the strike aircraft needed a warning system capable of evaluating accurately and immediately the nature of the threat which lay behind intercepted enemy radar emissions. Whenever a pilot penetrated enemy air space, he knew that sooner or later he would be 'illuminated' by a missile guidance radar; in the stressful circumstances of carrier-borne attack operations the task of interpreting warning lights and strange acoustic signals produced by on board EW devices had to be as rapid and simple as possible. The equipment also had to have maximum reliability, as any failure during combat meant the certain loss of the aircraft.

The Americans, therefore, set about producing a new generation of EW devices which were a great improvement over their predecessors. In particular, a whole range of more powerful airborne jammers were built, capable of totally jamming all types of enemy radar, and improvements were made to intercept receivers. Superheterodyne analysis receivers used in conjunction with automatic time/threat-visualisation correlation circuits proved to be particularly useful.

Over the next few years, US commanders and strategy were changed, while the frequency and intensity of Vietcong night assaults and ambushes were stepped-up. US bombing raids on Hanoi and Haiphong by B-52s became more frequent and more intense, but with equally frequent intervals of suspension.

The Russians provided North Vietnam with new SAM-2 missile batteries which featured a new version of the *Fansong* radar (there were at least seven versions altogether). The main difference between the first and the second version was that it operated on a higher frequency, in the 4910–4990 and 5010–5090 MHz bands, while the first type operated in 2965–2990 and 3025–3060 MHz bands.

A new type of Soviet ECCM was also introduced which used a highly original and effective new scanning technique in which a 'non-scanning' beam was used to illuminate the target, while the reflected signal was received by a scanning antenna which did not itself radiate electromagnetic energy. This ingenious technique, named LORO (Lobe-On-Receive Only), proved extremely effective and showed how far the Soviets had progressed in the field of electronic warfare. Once again, life became more difficult for US pilots in Vietnam. Meanwhile, air raids over Hanoi using B-52 bombers had been stepped up.

The North Vietnamese had also introduced an invention of their own, a simple but effective electronic 'trap' for the B-52 bombers. The US aircraft, based on the island of Guam in the middle of the Pacific Ocean, had little choice as to the route they could take to reach Hanoi or Haiphong and so, knowing their route, the North Vietnamese placed simple transmitters along the way to simulate the presence of *Fansong* radars. These would be turned on as US aircraft approached thus inducing the US pilots to launch their anti-radar missiles. This deception worked very well and US pilots frequently expended their whole load of ARMs on the false targets, leaving them vulnerable to attacks by real SAM-2 missiles over the target and on the return journey.

Immediately the Americans realised the shortcomings of the ECM equipment on their B-52s, particularly the jammers which proved unsuitable for missions carried out at such high altitudes, they set about modifying existing equipment and installing chaff-launching dispensers. However, the North Vietnamese countered this ECM by endowing their own radars with the ability to rapidly change frequency at the first sign of jamming: this ECCM is called 'frequency agility'.

Nevertheless, losses of B-52s throughout the LINEBACKER II operation, the bombing of Hanoi and Hoiphong, decreased considerably. In the course of 700 missions, about 1000 missiles were launched by the strong and sophisticated enemy anti-aircraft defence systems but only fifteen bombers were shot down, showing a loss-rate of 1.5 percent. US Air Force commands have estimated that, if the B-52s had not been fitted with all the most modern electronic warfare equipment, the number of aircraft shot down during these missions would not have been less than seventy-five. The highs and lows in the loss-rate of fighter-bombers also showed a close correlation with the advent of new arms systems and ECMs.

At this stage, US air power was the only way of exerting pressure on the North Vietnamese government to force them to negotiate and put an end to the fighting. On the ground, the Americans were on the defensive, and the war was already lost from a political point of view.

However, 1968 and 1969 were successful years for the North Vietnamese, not only on the ground but also in the air mainly due to massive reinforcement of their anti-aircraft defences. In one month alone, the Americans lost over ninety aircraft, most of them shot down during air raids over North Vietnam. Increasing numbers of aircraft were sent on such raids, reaching a peak of 400 aircraft on 1 May 1968.

From 1970 onwards, until the end of the war, air losses decreased progressively, mainly due to steps taken to facilitate penetration of enemy defences. Technological progress led to improvements in airborne EW devices, particularly RWRs in which digital techniques, hybrid micro-circuits and special microwave components were incorporated. In this period, the first computer-controlled RWR was produced with the ability to simultaneously and instantaneously analyse the parameters of all intercepted electromagnetic signals.

In 1971, a new aircraft entered service, the Grumman EA-6

Prowler, specially designed for electronic warfare. Besides RWR it was equipped with powerful jammers to jam enemy search radars while the fighter-bombers went in to attack enemy anti-aircraft batteries. During these missions, the jamming aircraft had to stay out of range of the missiles themselves and, for this reason, this type of jamming was named Stand-Off Jamming.

Finally, the Americans fitted their aircraft with a piece of equipment belonging to the new generation: the 'smart' or deception jammer. This was able to deceive an enemy radar by producing a false echo on the radar's screen, giving false information regarding the distance, direction and speed of the real echo. The result was that a missile was guided towards a non-existent target and away from the real one.

However, in spite of so many new, sophisticated inventions in the field of military technology, the Americans after having been involved directly or indirectly in that 'semi-war' of Vietnam for almost ten years, had to quickly and definitively withdraw from that troubled area of Southeast Asia.

There is no doubt that, during the whole war, the application of ECM led to a decrease in air losses. At the beginning of the war, the loss-rate was 14 percent whilst by the closing stage of the war it had dropped to 1.4 percent. This was not a steady decrease, however. Whenever the North Vietnamese came up with weapons controlled by new types of radar, US loss-rates rose, and whenever new EW equipment to counter these threats were installed on US aircraft, they began to fall again.

In the conflict between radar and ECM, and between ECM and ECCM, the dynamic nature of electronic warfare is thus clearly demonstrated.

17

The Arab-Israeli Wars

The Six-Day War

After the Sinai campaign of 1956, the second in a long series of short wars fought in the Middle East, there followed a fairly long period of calm in that troubled part of the world. During this period, both Arabs and Israelis set about modernising their armed forces on the basis of what they had learnt from the last conflict.

The Israelis received a number of Hawk ground-to-air missiles from the United States and Centurion tanks from Great Britain. France supplied them with Dassault-Breguet Mirage III and Super Mystère jet fighters which greatly improved the quality of the Israeli air force (Chel Ha 'Avir). The Israelis were now in possession of a highly efficient air force, equipped with excellent aircraft and helicopters.

Meanwhile Russia was supplying Egypt with a large number of modern weapons, such as MiG-21 fighters and Tupolev Tu-16 bombers.

In the first few months of 1967, relationships between Israel and neighbouring Arab countries began to deteriorate following a series of incidents along Israeli borders. In the spring of that year, Egypt asked the United Nations to withdraw neutral forces which had served as a buffer in the Sinai peninsula and sent 100,000 of her own troops and about 1000 tanks into the area. Tension reached a climax when the Egyptian President, Nasser, closed the Straits of Tiran, thus preventing Israeli ships from reaching the Red Sea from the Gulf of Aqaba. Israel immediately mobilised her troops in readiness for strategic action, which would have to be in the form of a lightning attack. The reasons for this operational necessity were not only that international intervention would put an end to hostilities as soon as it became apparent that Israel was winning but also because nationwide mobilisation would soon lead to the economic paralysis of the country.

The Arab nations (Egypt, Syria, Iraq and Jordan) deployed nearly a million well-equipped soldiers, 700 combat aircraft and over 2000 tanks along Israeli borders, ready to attack the enemy from all sides.

The whole world waited with bated breath, wondering whether they were on the brink of a third world war. The various electronic information-gathering systems of the major world powers were all focussed onto the situation in the Middle East.

The ships of the Soviet fleet in the Mediterranean, particularly those ships specially adapted for electronic espionage, were constantly tuned in to all frequencies of the electromagnetic spectrum to keep an eye on the delicate situation.

The US Sixth Fleet was cruising in the eastern waters of the Mediterranean and special aircraft equipped with the most advanced electronic devices kept the area of Israel, Sinai and practically the whole Middle East area under constant surveillance. The US Navy SIGINT ship USS *Liberty* was on continuous patrol off the coasts of the Middle East, just outside territorial waters. She was equipped with highly sensitive electronic equipment which could intercept and decipher all radio communications transmitted by both Arabs and Israelis and intercept and analyse all their radar emissions. The British were also keeping a close eye on the situation from their vantage-point of Mount Trudos in Cyprus.

Naturally, all Egyptian radar stations were on constant alert; there were twenty-three altogether, sixteen of which were situated in the Sinai peninsula. All air space and coastlines surrounding Egypt were covered by their early warning radar systems. The Israelis were also continuous radio and radar alert.

The Six-Day War broke out on 5 June 1967; this day also marked the beginning of a series of electronic challenges which was to continue for many years. At exactly 07.45, the Israelis launched a surprise attack intended to ensure their total air superiority which would in turn enable them to achieve their other objectives. Their very elaborate and precise plan of action was to attack enemy airbases and destroy all the enemy's combat aircraft whilst they were still on the ground. The *sine qua non* for the success of this action was that the enemy be taken by surprise and that all his communications and surveillance systems be paralysed. In order not to arouse the slightest suspicion of the imminent attack, the Israelis had also drawn up elaborate plans to deceive the enemy. Regular morning training flights were carried out as usual, the attack being scheduled for 07.45 when the Egyptian pilots, having been on the alert from dawn to 07.30—the period in which attacks are normally launched and wars normally break out— would be going to have breakfast in the airbase canteens and staff officers would be going to their offices.

Previous aerial reconnaissance flights had shown the exact deployment of enemy air squadrons and radar stations, radar coverage and blind spots; the Israelis had even managed to plot a route through the towers and minarets of Cairo whereby, flying at such a low altitude, they would be masked from enemy radar and could launch a surprise attack on the West Cairo airbase which housed not only the MiG-21s used for defence of the capital city but also the huge Tu-16 bombers which would be used for air raids on Tel Aviv. By skilful interpretation of reconnaissance photographs, the Israelis had managed to distinguish the real aircraft from the many dummy aircraft which the Egyptians had set up to deceive enemy pilots.

Instead of flying directly towards their targets, the first wave of Israeli aircraft flew out to sea off the Egyptian coast, swung round and, flying low just above the water, approached from the West, exactly the opposite direction to which the Egyptians would expect them.

The attack took place at exactly 07.45 as scheduled. At that time, every Egyptian aircraft was on the ground except one, a twin-engined Ilyushin which was flying towards the Israeli border with three of the highest-ranking officers of the Egyptian armed forces on board. One of them was the Chief of Defence Staff, General Amer. They were listening in on communications transmitted on the frequency used by Israeli pilots during their normal flights but nothing unusual had been intercepted on that frequency. Suddenly, the control tower of an Egyptian military airbase communicated to the generals that the base was under enemy air attack. It was exactly 07.45; the Ilyushin turned back immediately and, while they were heading back to base, the generals radioed to ground commands to try to find out what was happening, but all they could hear was a babble of voices and other noises. Every time they tried to land somewhere, they realised from the few clear words they managed to pick up that the base was under attack. They tried several times to land at one of the many Egyptian air bases on the Suez Canal but they were all under attack and their runways had been put out of action. Finally, the Ilyushin managed to land at Cairo International Airport and the three generals rushed to high command headquarters where they were informed that practically the entire Egyptian air force had been wiped out.

Given the limited number of aircraft at their disposal and the fact that the distance they had to cover was fairly short, the Israelis were able to send each aircraft out several times, thus multiplying the number of missions that could be accomplished. After the initial attack, which came as a complete surprise to the Egyptians because their radar and radio communications had been blanked out, the Israel aircraft returned to base to be re-fuelled and re-armed and were then sent out again with new pilots. After having destroyed 300 of the Egyptian air force's 320 aircraft, the Israelis immediately went on to destroy the air forces of the other Arab states bordering Israel. In quick sucession, the Jordanian, Iraqi and Syrian air forces were wiped out.

In less than two days and with a fairly small number of aircraft at their disposal, the Israeli air force flew about 1100 missions with many pilots flying eight to ten missions a day.

Now that they had gained absolute air superiority, the Israeli air force could be committed to tactical air support for ground forces. The Egyptian expeditionary corps in the Sinai, made up of 100,000 men, was completely overrun by Israeli armoured columns and fell back in disarray, abandoning many brand-new weapons systems, including tanks and electronic equipment only recently received from Russia.

The Egyptian President, Nasser, could not believe that the Israelis had achieved such outstanding results in such a short time without outside help. Arguing that it was impossible for such a small air force to accomplish so many missions, he tried to convince his ally King Hussein of Jordan that the Israelis had been helped by aircraft from the United States and Great Britain. However, the Israelis, who had been systematically intercepting all enemy electromagnetic emissions, made public their recording of a radio-telephone conversation which they held had taken place at 04.45 on 6 June between President Nasser and King Hussein. From this conversation, it was quite clear even to the Russians, that the two Arab leaders were plotting to spread the rumour that the USA and Great Britain had participated in the Israeli attacks. As a result of the interception of this conversation, international tension was diffused and the prospect of a war involving major world powers was avoided.

As always happens in wars, even the victors make mistakes and, in this war, the Israelis made several, one of which was extremely serious. On the afternoon of 8 June the US Navy spy-ship *Liberty* was attacked by two Israeli Mirage fighters and three gun-boats near El Arish. They had mistaken the ship for an Egyptian navy destroyer equipped with EW equipment for jamming Israeli radars. The unfortunate ship was badly damaged and thirty-four of the crew were killed and seventy-five injured. The US government accepted Israeli explanations and apologies, although it was not really clear how the Israelis could have mistaken a spy-ship for a destroyer. Lt James M. Ennes, electronics officer aboard USS *Liberty*, stated that the attack had been too well-coordinated to have been a mistake. On the other hand, the US

Department of State's explanation of what the spy-ship was doing near Egyptian territorial waters was not very clear either. The ship was said to have been there to 'ensure' communications between American listening posts in the Middle East.

The truth of the matter is that USS *Liberty* and similar Soviet SIGINT ships were stationed in the area to intercept and record radio communications and radar emissions and retransmit them to their respective governments, both of which were keen to keep an eye on developments in that explosive part of the world.

How had the Israelis managed to destroy the whole Egyptian air force in the space of two hours, giving the Egyptians no time to react?

For obvious reasons of secrecy, the Israelis have never revealed their electronic plan of action. Nevertheless, considering the twenty-three Egyptian radar stations and numerous US and Soviet spy-ships in the area, it is difficult to accept that all the operators were asleep at the time of the attack. It is also difficult to accept that no orders were given to Egyptian pilots before and after each air raid.

The explanation lies in the fact that, in 1967, ECM was for the Israelis more than a mere memory of actions taken by the British during World War Two, when they transmitted false information, distorted radar signals used to guide enemy bombers and jammed enemy radars. The Vietnam war had been going on for some years and the Israelis knew that the Americans had resorted to electronic jammers to deal with Soviet SAM-2 missiles and radar-controlled 57 mm anti-aircraft guns.

No electronic warfare action was taken by the Israelis against Egyptian radar until 07.45 on 5 June, because the surprise element was essential for the success of the operation and, therefore, the Israelis could not risk arousing the enemy's suspicions. At 07.45, the most far-off radars were attacked and put out of action while those within range of Israeli electronic equipment were subjected to jamming. Moreover, during and after the initial attack, Israeli radio operators who spoke fluent Egyptian Arabic transmitted into the enemy air defence radio communications network, giving false orders, cancelling correct orders and generally causing confusion and preventing Egyptian commands from using the radio. The Israelis also jammed Russian and American radar and radio communications, in some cases.

The 'War of Attrition'

The Israelis naturally expected that their victory in the lightning Six-

Day War would ensure a long period of peace and allow them to negotiate a lasting peace from a more favourable position.

This conviction was supported by all the results of their short but successful campaign: the entire Egyptian war machine had been destroyed, Jordan had lost most of its army and some of its territory and Syria, too, had lost important military positions, such as the Golan Heights. Most important of all, the Israelis now occupied the Sinai peninsula which would be a comforting buffer-zone between themselves and a belligerent Egypt. Here they could set up a network of warning radars and other sensitive electronic systems to maintain surveillance of the hostile region, something they had wanted for a long time.

However, this ideal state of affairs was not to be. The end of the Six-Day War marked the beginning of a long series of military actions by both sides—the so-called 'War of Attrition'—made possible by the increasingly sophisticated electronic warfare equipment possessed by both sides.

The Soviets, fearing that the Suez Canal might fall into Israeli hands, quickly replenished the depleted Egyptian war machine by supplying aircraft, tanks and modern artillery to discourage any attempt by the Israelis to cross the Canal. Only two weeks after their defeat, the Egyptians received from Russia 200 aircraft, mainly MiG-21s and Sukhoi Su7s, modern T-55 tanks and a number of radar-controlled anti-aircraft guns. Egypt thus replaced 70 percent of her losses and now possessed weapons of much higher quality than before the war.

Since, during the Six-Day War, most Egyptian aircraft had been destroyed on the ground, very few pilots had been lost and so the Egyptian air force was able to send many of them on training courses in Russia without unduly depleting their operational force.

However, for the Israelis, the loss of forty aircraft in combat was a major loss considering the limited size of their air force. For political reasons, Israel's usual supplier, France, had placed an embargo on the delivery of a further fifty Mirages and the United States was reluctant to supply the A-4 Skyhawk fighter-bombers that they had promised.

Peace negotiations were shelved as the Egyptians continued to pile up new weapons. Egypt refused to accept the new territorial situation, in favour of the Israelis, and began sporadic bombing of Israeli positions along the East bank of the Suez Canal.

On 21 October a few months after the Six-Day War, an incident took place which greatly boosted Egyptian morale: the 1710-ton

Israeli destroyer *Eilat*, the ex–British World War Two *Zealous*, was hit and sunk by missiles launched from two Egyptian torpedo-boats anchored in Port Said. Of the 202 men on board the *Eilat*, forty-seven were killed and ninety-one were injured. The incident caused great consternation because it was the first time that a warship had been sunk by missiles and also because there was little in common between the two versions of what had actually happened.

According to the Israelis, the *Eilat* had been at a distance of 14 miles from Port Said, 2 miles outside Egyptian territorial waters, when she had been hit by two Styx missiles launched from Soviet-built 'Komar' class patrol boats. The Soviet-built missiles had radar-guidance, a range of 25–30 miles and an explosive charge of 880 lbs. On being hit, the Israeli ship had keeled over but had not gone down. Two hours later, the Egyptians, seeing that the ship was still afloat, launched another two Styx missiles, one of which sent the ship down while the other exploded in the water, killing or injuring many crew members.

According to the Egyptians, the *Eilat* had been at a distance of only 10 miles from Port Said, within Egyptian territorial waters. Only two missiles had been launched but these had been sufficient to sink the Israeli ship immediately. The Egyptians also denied allegations that they had had Soviet advisors on board their patrol boats.

Whatever the case, the sinking of the *Eilat* was a great victory for the Egyptians, strengthening their faith in their armed forces and their determination not to discuss peace with Israel.

From a more objective point of view, the sinking of that destroyer by missiles launched from small patrol boats had a strong effect on naval thinking and marked the beginning of many changes in the design of warships and their weaponry, not to mention the tactics governing their use. This incident was a rude awakening for all the world's major navies, forcing them to realise that even their largest warships were practically defenceless against this new threat from missiles which had a greater range than naval guns and, moreover, could be launched from small fast boats like torpedo-boats or patrol boats. But what caused the greatest consternation among major navies of the West was the fact that Russia had supplied several minor navies, some from communist bloc countries, not only with 'Komar' class fast attack boats but also with a number of larger 'OSA'-class boats which could launch more missiles than 'Komar' class vessels.

Israel's reponse to the sinking of the *Eilat* was quick and violent: on the afternoon of 24 October, the Israelis first bombed the city and port of Suez and then attacked two large petroleum refineries, situated in the coastal zone, which produced about five million tons of fuel annually. Israeli aircraft also attacked the base in the port of Alexandria where 'OSA' and 'Komar' missile boats were anchored.

After these attacks, the banks of the Suez Canal became the scene of frequent fighting with daily artillery duels, commando raids, air battles and bombing attacks, particularly by the Israelis who had local air superiority.

Not having fire-control radar, Egyptian anti-aircraft artillery was unable to cope with Israeli air raids. To redress the balance, at the end of 1968, the Russians decided to provide Egypt with SAM-2 missile systems; these had made their first appearance in 1965 in Vietnam and had been used towards the end of the Six-Day War on the Syrian Front. They were deployed in a 16-mile wide strip along the West Bank of the Suez Canal. However, as in Vietnam, the SAM-2 system did not have great success due to intrinsic limitations (mentioned in chapter 16 on the Vietnam war). Besides these shortcomings, two other defects were revealed in the war theatre of the Suez Canal: the first was the limited mobility of the system, which had to be towed on a trailer and required time to set up before going into action; the second was that the radar-guidance system only worked above 6000 metres and was therefore ineffective against low flying aircraft.

Like American pilots in Vietnam, the Israelis had also learnt to recognise the famous 'SAM-song', the characteristic sound of the SAM-2 radar pulses which meant that a missile was heading towards their aircraft. It would seem that the Israelis had captured a SAM-2 system in 1967 during the last phase of the war fought in the Golan Heights in Syria; there is certainly no doubt that they knew its precise operating frequencies since they had already devised appropriate jammers which could only have been done on the basis of such knowledge.

Thus an 'undeclared' war had broken out in the Middle East, an electronic war fought 'by proxy' since neither Egypt nor Israel had an industry capable of producing such technologically advanced electronic systems as were necessary and, therefore, used equipment supplied by the Soviet Union and Western powers respectively.

In order to devise ECMs to counter radar guided missiles effectively, it is first necessary to know the precise characteristics of the radar used. Therefore, in 1969, both the Israelis and the Egyptians started to make raids on enemy territory whose aims were to capture radar sets in which they were interested, or, at least, one of the set's main components, often enough to yield the information sought.

In June 1969, the Egyptians made three raids in which they managed to destroy several Israeli radar installations. In the same month, the Israelis also made a raid in an area about 6 miles south of Suez in which they claimed to have destroyed an enemy radar installation. Again in June, the Israelis announced that they had captured an Egyptian coast guard boat in the Gulf of Suez. In July, an Israeli Commando attacked a fortification on Green Island in the northern part of the Gulf of Suez. The tower where the radar was installed was surrounded by high walls which the Israelis scaled and, after killing the guards, removed the desired parts of the radar and destroyed the rest. The whole operation took about one hour.

On 9 September, the Israelis organised a full-scale military operation on the south coast of Egypt. At dawn, a small convoy of gun-boats and landing craft carrying six tanks and three armoured cars set sail from somewhere on the Sinai coast and headed for the south coast of Egypt. The vehicles were all Soviet-built and had all been captured during the Six-Day War and still bore markings which showed they belonged to the Egyptian Army. The 150 Israelis on board wore Egyptian uniforms and spoke perfect Arabic. Just after daybreak, the disguised Israeli commandos landed about 30 miles south of Suez and, undetected, headed straight for the radar installation near El Khafayer, which they quickly dismantled. They then headed south along the coast road where they were joined by air and naval escorts which helped them to destroy all enemy military installations along the route.

Radar installations near Ras Darg were the first to be destroyed, followed by those near the small port of Ras Zofarana, about 56 miles south of Suez; here, they captured some new Soviet-built armoured vehicles which they took back to Israel, along with all the captured electronic equipment, for further examination.

However, the most significant episode, from an electronic point of view, was the Israeli capture of a complete P-12 *Spoon Rest* radar set from the Egyptian naval base of Ras Ghaleb on the Red Sea in a raid which took place on 27 December 1969. The Soviet-built radar had recently been installed at Ras Ghaleb to complete Egyptian early warning radar coverage. With a range of 270 kms it was able to detect an aircraft taking off from any Israeli airbase on the other side of the Suez Canal and track it until it was within range of a *Fansong* SAM-2 radar. The two radars, *Spoon Rest* and *Fansong*, worked in conjunc-tion; at a certain point, the aircraft would be turned over to the *Fansong* radar which would guide a missile towards it.

The *Spoon Rest* radar weighed seven tons and needed two large trucks to move it. The Israelis knew practically nothing about its electronic characteristics and so, if they wanted to devise effective ECM, they had no alternative but to capture a set and examine it. This was the purpose of the commando raid at Ras Ghaleb, about 115 miles south of Suez.

To distract attention from this operation, an air raid on the Egyptian side of the Canal was scheduled to take place at the same time. After landing, the Israeli commando skirted the radar installation and, going via the desert, came in for attack from the landward side. After they had occupied the base, two large helicopters were flown in to pick up the radar and take it back to Israel.

Having an actual example of the enemy radar set in their possession greatly facilitated the Israeli's task of devising ECMs to jam or deceive radars of that type. They were able to discover the exact operating frequency and other important features, including ECCMs incorporated into the set to protect it from enemy ECMs.

Israel was not alone in its anxiety to find ECMs capable of impairing the effectiveness of the *Spoon Rest* radar. The countries of the NATO alliance were also very interested and it was not long before Western electronics experts came up with appropriate devices for jamming and deception.

Towards the end of 1969, the Israelis again hit the newspaper headlines by removing five 250-ton fast patrol boats from France. These were the last of twelve boats ordered by Israel before the embargo decreed by President de Gaulle. The five vessels were being held at the port of Cherbourg with a limited number of crew members on board each. On Christmas night, taking advantage of laxer security during the festivities, the vessels slid out of port and, once out to sea, the Israeli crews made full speed ahead to their homeland, arriving on New Year's Eve at Haifa where they were eagerly greeted by a joyous crowd. The five gun-boats, like the other seven already in Israel, were equipped with surface-to-surface Gabriel missiles built in Israel as well as with active ECM equipment.

Meanwhile, the undeclared 'war of attrition' along the banks of the Suez Canal was becoming more and more intense, each commando raid being followed by a reprisal from the other side. Israeli aircraft began to be used as flying artillery against SAM-2 missile ranges along the banks of the Canal. Many Egyptian missile systems were hit but many Israeli aircraft were brought down.

Assistance for the Israelis soon arrived from the United States, which supplied McDonnell-Douglas F-4 Phantoms. The F-4 had played a leading role in air battles over Vietnam. They were excellent in the role of interceptor but were equally suited to providing tactical close air support for ground forces and making ground attacks. But perhaps the most important thing was that, along with the F-4 Phantoms, the Israelis received electronic warfare equipment, housed in special pods fitted externally to the aircraft.

Pods containing RWR were also installed on the A-4 Skyhawk light tactical fighters; the pods also housed new jammers capable of totally blanking out *Spoon Rest* radars, based on the research done on the *Spoon Rest* captured some months before. These new jammers, and other devices capable of surveying the electronic situation in the entire Suez area, were also installed on several modified Boeing B-47 Stratocruisers. The B-47 had been one of the first aircraft to be dedicated exclusively to electronic warfare. These aircraft proved to be of great value to the Israelis. At high altitudes and, for obvious reasons of security, far away from enemy lines, they flew parallel to the Suez Canal and the Syrian border and were able to monitor enemy air activity and, when necessary, paralyse their air defence radars.

In the first three to four months of 1970, Israeli aircraft, protected by this electronic shield, were able to penetrate more deeply into enemy territory. Their first target was the enemy radar network along the Canal; next, it was the turn of the inland air defences comprising the search and fire-control radars deployed around the Egyptian capital, the Aswan Dam and other important installations.

The Egyptians retaliated with repeated artillery shelling along the Canal and air raids, using MiG-21s, which penetrated deeper into Israeli territory. However, these were not as successful as the Israeli air raids which reached Cairo and beyond, heedless of SAM-2 missiles, thanks to their new EW equipment.

Egypt asked the Soviet Union for more effective arms and equipment to deal with Israeli air offensives. In the spring of 1970, the Soviets duly furnished Egypt with SAM-3 missile systems, NATO code-name *Goa*. The SAM-3 had a range of 34 kms and much greater mobility than SAM-2, being mounted on ordinary vehicles and was also effective against low flying aircraft (300–45,000 feet). Each SAM-3 system had four missile-launchers which worked in conjunction with two radars; a search radar (NATO code name, *Flat Face*) and an acquisition radar (NATO code-name, *Long Track*). The former had the task of detecting the intruders while the latter tracked intruders once acquired and tracked them with sufficient accuracy to enable missiles to be launched at them.

Russia also delivered a new version of the MiG-21, the MiG-21J which was equipped with a new, more sophisticated type of radar and, compared to earlier models, had a superior endurance enabling it to operate deep into Israeli territory.

Ever increasing numbers of Soviet technicians, instructors and pilots accompanied the new equipment, and, after a few months, the Soviets assumed control of Egyptian air defence organization. The Israelis soon realised, through their recordings of Egyptian flight communications, that many MiG-21Js were flown by Russians and wondered with some consternation what would happen if one of them were shot down, an event which was sooner or later bound to happen.

The employment of Soviet pilots, although limited to the Cairo zone and other important inland locations, greatly helped the Egyptian air force, freeing the Egyptian pilots to concentrate on offensive actions and reprisal raids against Israel. Both sides suffered heavy losses in the frequent air battles over the Suez Canal although reports about them seldom tallied. Meanwhile, both sides continued to receive new, sophisticated arms and electronic equipment from the Great Powers who seemed to think the Middle East was one huge missile range where they could try out their new weapons systems under real tactical conditions.

The Israelis received an electronic apparatus which can be considered the most secret of all electronic warfare equipment: a deception jammer. This is an electronic device capable of falsifying data regarding distance, direction and speed which an enemy missile-guidance or fire-control radar is trying to acquire. If a missile is heading towards a target (land, air or sea) equipped with a deception jammer, signals are produced in the missile-guidance radar by the deception jammer which show the target in a different position to its actual position. Thus the missile, instead of continuing on course towards the real target, is sent off course by this misleading information.

The remarkable advantage of this process is that the enemy radar does not become aware of the deception because the return echo from the target is always exactly what is expected. This is possible because the distance of a target is calculated by measuring the time-lapse between the transmission of an electromagnetic pulse and its return echo. When the deception jammer-equipped target aircraft or ship is illuminated by the enemy radar, it is sufficient to simply delay the return echo or modify its width in order for a wrong distance or

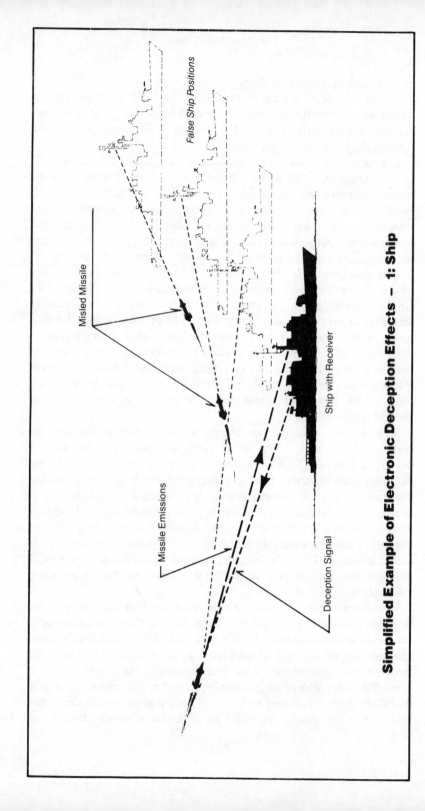

Simplified Example of Electronic Deception Effects — 1: Ship

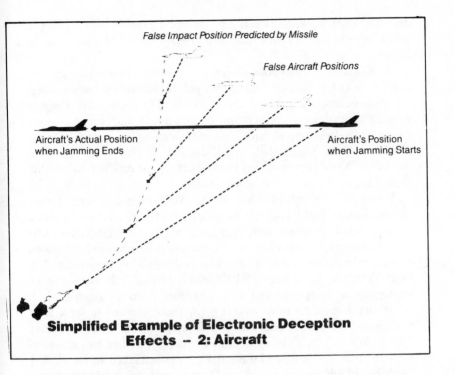

False Impact Position Predicted by Missile

False Aircraft Positions

Aircraft's Actual Position
when Jamming Ends

Aircraft's Position
when Jamming Starts

**Simplified Example of Electronic Deception
Effects -- 2: Aircraft**

Examples of electronic deception. A repeater (or deception) jammer, installed on a ship or an aircraft, 'captures' the signals from the seeker head of a radar-guided missile, or an AA battery. The signals are momentarily delayed and altered before being retransmitted, giving a false radar echo. The missile thus receives incorrect information about the target's range, course and speed, and, consequently, its position, and guides itself on the false echo, due to the incorrect information, so missing the target.

direction to be shown on the enemy radar screen.

Deception jammers are quite small and can easily be installed on aircraft, either internally or externally in pods attached to the same hard points as those used for bombs or external fuel tanks. From an industrial point of view, the manufacture of a deception jammer requires very sophisticated technical capacity and an extremely advanced technology which is not so easy to acquire.

The arrival of these new electronic gadgets was a real morale booster for the Israeli forces and General Moshe Dayan predicted a hot 'electronic summer', which promptly arrived. In June 1970, a dramatic duel between Israeli aircraft and Egyptian missiles began which resulted in the destruction of nearly all SAM-2 systems in Egypt.

At this point, the Egyptian air force received the first of the eagerly-awaited Soviet MiG-23s: these ultra-modern, multi-role 'swing-wing' (variable-geometry) fighters, given the NATO code-name *Flogger*, were equally adept at interception, ground-attack and reconnaissance and were well-equipped with new radar-controlled missiles for air combat. Being faster than the Israeli F-4 Phantoms and A-4 Skyhawks, their appearance in the skies of Egypt produced a notable slackening in the pace of Israeli air raids.

Above all, the introduction of the MiG-23 meant that Israeli reconnaissance flights had to be reduced since the aircraft used for this purpose were unarmed and, therefore, extremely vulnerable. The Israelis were thus deprived of the information which is indispensable in modern warfare. To overcome this problem, they resorted to US built Teledyne Ryan 124-1 RPVs which carried only electronic or photographic equipment and were controlled from the ground.

During Israeli air raids over Egypt, there seemed to be a tacit understanding between Israeli and Russian pilots to avoid direct conflict at all costs. This lasted until 25 July 1970 when two Russian-piloted MiGs suddenly and quite deliberately attacked an Israeli A-4 Skyhawk which, however, managed to get away. After this encounter, the Israelis had no option but to abandon all precautions. Thus, on 30 July, when a squadron of Israeli Phantoms was intercepted and attacked by sixteen Russian-piloted MiGs, Israeli Mirage fighters came to the rescue and, after a few minutes of fierce combat, five MiGs had been shot down. The Israelis also suffered losses, albeit undeclared: three Phantoms did not return to base that day and were probably shot down by Egyptian anti-aircraft artillery on their way home.

By now both the Egyptians and the Israelis, who were fighting a sort of 'proxy' war for the two superpowers, realised that this war was no longer worth fighting because they risked provoking a general war which neither side desired at that particular moment. Consequently, on 7 August 1970, they accepted, without too much argument, a ceasefire proposed by the USA which put an end to almost three years of inconclusive and bloody fighting. Both sides suffered heavy losses, although these were officially either not admitted or minimised for propaganda reasons: it is, therefore, difficult to give exact figures but it can be reasonably estimated that the Israelis incurred casualties of no less than 400 dead and about 4000 injured while the Arabs suffered casualties of about 1500 dead with about 7000 injured. Estimates of air losses are probably more accurate as there was some agreement between different sources: about 105 Egyptian and sixteen Israeli aircraft were shot down, of which only seven were brought down by missiles.

This great discrepancy in favour of the Israelis can be largely attributed to the EW equipment installed on board their aircraft which, in the duels between missiles and aircraft, saved the lives of many Israeli pilots.

The Yom Kippur War and the Technological Surprises

On 6 October 1973, the Jewish Day of Atonement or Yom Kippur, the Arabs made a surprise attack of unprecedented violence while almost the entire population of Israel was praying. At 14.00 Egyptian Su-7s and MiGs began to attack Israeli defences and airbases in the Sinai while about 4000 guns of various calibres began a massive barrage of the Bar-Lev defence line and other important installations on the Suez Canal.

The Arabs then started jamming Israeli radio communications, making it impossible for the Israelis to exchange battle orders. In addition, some Israeli radio and radar stations along the Canal were destroyed by special teams of Egyptian scuba divers.

On the Syrian Front, Soviet-built Sukhoi aircraft of the Syrian air force swooped down on the Golan Heights and destroyed nearly all Israeli defence installations in that area.

A few minutes later an avalanche of over 800 Russian-built Egyptian tanks swept across the Suez Canal, crossing it at many points by mobile pontoon bridges which had been set up in record time. The Israelis were caught completely off-guard and, since many troops had been given home leave for the holiday, defences were greatly reduced. The Bar-Lev defence line was completely overwhelmed by the mass of tanks advancing across the Canal. Israeli napalm defences, which should have set fire to the whole Canal zone, had been carefully disconnected by Egyptian scuba divers who had worked undetected for several nights before the attack.

After a few hours of total confusion, Israeli high commands managed to rapidly sketch together a defence plan. The air force reacted first, sending Phantoms and Skyhawks to attack. The Israelis were confident of the superiority of these aircraft which, mainly due to the sophisticated EW equipment they carried on board, had already demonstrated their superiority in battle with the enemy. However, their confrontation with the advancing Egyptian armoured columns was nothing less than a disaster. The Israeli pilots did not hear the usual 'SAM-song' and could therefore do nothing to avoid enemy missiles. In the first two to three days of the war, a great number of Israeli aircraft were shot down.

Obviously something had changed in the electromagnetic spectrum as the electronic devices installed on Israeli aircraft were no longer effective. A first appraisal of the situation showed that the radars used to guide Egyptian missiles and gunfire were operating on a higher frequency and using more sophisticated guidance techniques than the SAM-2 and SAM-3 missile systems.

The Israeli pilots who survived that veritable massacre of Phantoms and Skyhawks reported that the advancing enemy columns were protected by an extremely effective and varied mobile anti-aircraft defence system. First of all, there was a screen of ultra-modern SAM-6 *Gainful* missile systems mounted on armoured vehicles; next, came the four barrel 23 mm radar-controlled ZSU-23-4 *Shilka* anti-aircraft guns mounted on tank-chassis; finally, there were the light man-portable, shoulder launched SAM-7 *Strela* infrared missiles, for low-level AA defence. Together these formed an almost impenetrable air defence ring, a mobile umbrella under which the armoured tanks could advance in safety, sheltered from air attacks.

The real strength of this system lay, not in its firepower or other factors of that nature, but solely in its weapon-guiding systems, which constituted a great technological surprise not only for the Israelis but for all the Western powers.

The SAM-6 system, whose main task was to provide anti-aircraft defence for field forces, comprised two tracked vehicles, one of which carried three *Gainful* missiles and the other the radar, code-named *Straight Flush* by NATO. The novelty of the system was that it used continuous waves (CW), unlike the SAM-2 and SAM-3 systems which used pulse waves. A target was illuminated by a low-power CW signal emitted by the *Straight Flush* radar, and the SAM-6 missile would home on to it by following the reflected energy. Since the receivers on board Israeli aircraft were designed to intercept pulsed signals, these CW emissions were not picked up. To make things even more difficult, the *Straight Flush* radar operated on two different frequencies. Thus, as a result of these two technological innovations, the *Guideline* missile could now approach the enemy aircraft without being discovered, jammed or deceived by the Israeli electronic devices.

Another technological surprise was the *Gun Dish* radar used to control the mobile 23 mm ZSU-23-4 *Shilka* AA guns. To avoid enemy ECM, this radar used a much higher frequency than any previously used by the Egyptians. The Israeli receivers, which were built to intercept frequencies of up to 12,000 MHz (12 GHZ), were unable to reach the electromagnetic emissions of the *Gun Dish* radar which had a frequency of about 16,000 MHz (16 GHz).

Another innovation was the small *Strela* anti-aircraft missile which a soldier could carry on his back. This had a completely new kind of guidance-system, using infrared (IR) rays. All the soldier had to do was aim the missile in the direction of a low-flying enemy aircraft. The missile's infrared detector would detect their heat emissions from the aircraft's jet engines, passing signals giving range and bearing to the control and guidance system which would guide the missile onto the target. Such a missile guidance system is termed IR-homing.

These new weapons systems, together with those already in existence (SAM-2 and SAM-3), constituted a truly exceptional air defence system, permitting the Egyptians to advance even though their air force had not achieved air superiority. Israeli aircraft, committed to battle in support of their ground forces, by attacking enemy armoured columns, found that there was no way to avoid that network of fire; if they dived to low altitude to avoid SAMs, they inevitably flew straight into vicious flak from the rapid-firing *Shilka* guns, or became targets for small *Strela* missiles. Israeli air losses were so high that ground commands decided no longer to request air support against enemy armoured columns.

The situation was hourly becoming more critical for the Israelis on both fronts as, in addition to the huge number of aircraft lost during the first days of the war, their tanks had also been massacred, easy targets for the new Russian-built *Sagger* anti-tank missiles. Launched by infantry at close range, these wire-guided missiles were extremely accurate.

Having by now realised that their nation was in serious danger of being overrun, the Israeli high command had to make the extremely important decision as to which front would be given defence priority. They judged the greater danger to be on the Northen front and, therefore, decided to concentrate on blocking the Syrian advance while trying to fend off Egyptian attacks in the Canal zone. The only hope for the air force, however, taken completely by surprise from the

EW point of view, was to try as quickly as possible to come up with effective electronic and infrared countermeasures, and thus reduce the absolutely unacceptable loss rate.

In those very first, dramatic days of the war, help arrived for the Israeli air force in the form of large quantities of chaff and chaff dispensers. Chaff was, of course, nothing new, having been used extensively in World War Two and Vietnam: the only alterations to the system was the adjustment of the length of individual strips to the frequencies of the new radars to be jammed. The chaff was contained in capsules in turn contained in pods attached to an aircraft, and was launched on command from an aircraft's pilot.

Besides chaff, the Israelis also received IR flares to deceive IR-guided missiles. These were used in the same way as chaff except that the flares generated heat, or IR energy. To achieve the purpose for which they were intended, the energy released had to be of the same frequency as that generated by the jet pipes of the aircraft's engine, but, obviously, had to be much more intense in order to create a false target towards which the SAM-7 missile could be guided.

As soon as the chaff and IR-flare-launchers had been installed on Phantoms and Skyhawks, the Israelis were able to devise tactics which

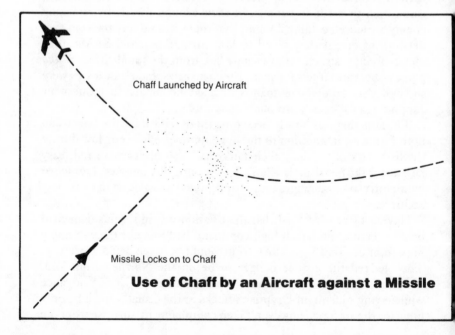

Chaff Launched by Aircraft

Missile Locks on to Chaff

Use of Chaff by an Aircraft against a Missile

An American-built AEW Grumman E2-C Hawkeye of the Israeli air force, used for stand-off electronic reconnaissance and intercept control.

The Cossor Electronics IFF 880/1, an example of the compact, light-weight equipment in use in the 1980s contained in standardised case sizes. An 'add-on' unit, rather than an integral part of the weapon system, it comprises a transmitter/receiver processing unit, carried in a haversack, and a hand-held open array antenna with an optical aiming system, and weighs only 6.5 kgs. It is compatible with a wide range of gun and missile air defence systems, typically Redeye, Stinger and SA-7. The AD weapon site commander sights the target through the antenna, and interrogates it using the trigger. A light indicates if the target is friendly or not. (Cossor Electronics via M G Burns)

An FFL pod-mounted Texas Instruments RS 700 series Infrared Line Scanner (IRLS) on a Danish air force Saab Draken. IRLS bridges the 'gap' between visual (camera) and radar reconnaissance to gather a far wider range of information. (Texas Instruments via M G Burns)

Infrared goggles for night vision worn by a soldier. These produce a half-tone image, enabling the wearer to pick out organic and man-made items against the natural background. Night-fighting is an extremely important part of NATO training, such techniques being used by the British during the 1982 Falklands Conflict.

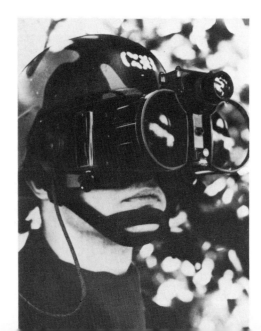

Two US Navy LTV A-7 Corsair II attack aircraft carrying pod-mounted Forward
Looking Infrared (FLIR) systems made by Texas Instruments. Such systems
enable all-weather operation on reconnaissance, gathering images when visual
photography is impossible, and on attack missions in low visibility or at night,
when a 'half-tone' image of the outside-world is presented on the pilot's Head-Up
Display (HUD). (Texas Instruments via M G Burns)

A Paveway laser-guided bomb (LGB), produced by Texas Instruments, mounted on a Royal Australian Air Force Dassault Mirage attack aircraft. Based on the Mark 82 bomb, a new tail-section and a laser-guidance nose-section are added. (Texas Instruments via M G Burns)

A Matra Otomat long range anti-ship missile, operational since 1977, being fired from the fast escort *Le Basque*. (MATRA via M G Burns)

Right: A Matra Otomat long range anti-ship missile being launched, for the first time, from a coastal battery at a target more than 100 kilometres away over the radar horizon. Within 6 minutes of leaving its container, the missile will impact. (MATRA via M G Burns)

Opposite: A Matra laser-guided bomb system mounted on a Dassault Mirage's wing pylon. LGBs, or 'smart bombs', provide staggeringly high accuracy compared to conventional bombing, and increase the survivability of the aircraft in the dangerous low-level attack area because it does not need to be exposed for so long while aiming. (MATRA via M G Burns)

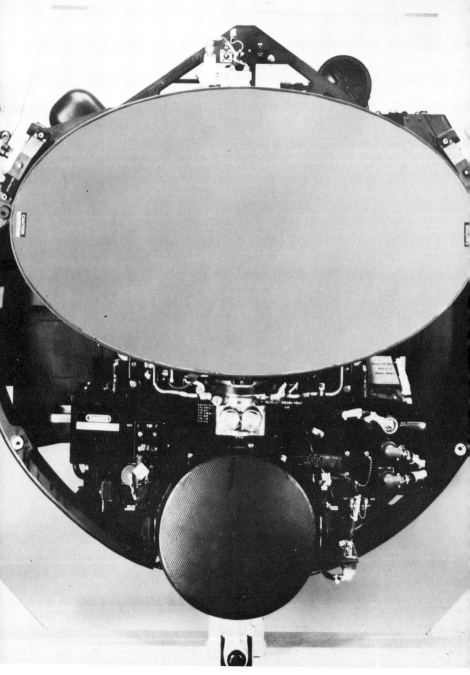

The multi-role radar by Texas Instruments used on the strike/attack Panavia Tornado in service with West Germany, Italy and the Royal Air Force. The elliptical scanner is for terrain-following radar, the smaller one for ground-mapping radar. (Texas Instruments via M G Burns)

Blindfire tracking radar with BAe Rapiers low level surface-to-air guided weapons system fire units. Rapier provides in depth point air defence, employing a direct hitting missile.

A French air force Dassault Mirage F1 interceptor firing a Matra Super 530 air-to-air missile. (MATRA via M G Burns)

A Matra Super 530 air-to-air missile carried on a launch rail on the underwing pylon of a French Dassault Mirage F1. (MATRA via M G Burns)

A USN F-4J Phantom II destroys an unmanned target aircraft over the Pacific Missile Test Center, Point Mugu, Ca., USA, with a Sky Flash. A semi-active radar-guided medium range boost and coast missile designed by BAe to replace the AIM-7 on the Phantom and Tornado F.2, Sky Flash has 'snap-down snap-up' capability, high lethality and considerable resistance to ECM. (Marconi via M G Burns)

Argentinian Super Etendards flying at very low level. Inset: An AM-38 Exocet stand-off air-to-surface anti-ship missile on a Super Etendard's wing pylon.

A Corvus eight-barrelled chaff and flare rocket launching system installed on the Royal Navy's medium and large warships. Each launcher fires 22-kg, 1130 mm long rockets. Although its principle function is chaff dispensing, it can also fire other types of decoy, such as IR flares.

HMS 'Sheffield' immediately after the Exocet attack. Inset: an enlargement (ringed B) of the UAA-1 Abbey Hill direction finding antenna.

HMS Sheffield during rescue operations with a Sea King hovering over her stern and HMS Amazon hove to alongside. The letters indicate: A Communications ESM antenna; B UAA-1 Abbey Hill antenna; C ECM antenna; D chaff launcher.

Soviet SAM-9 *Gaskin* surface-to-air missile system, mounted on 6 × 6 armoured cars for full mobility, seen on display during a Red Square display.

The Ford Aerospace Sergeant York US Army Division Air Defense Gun System, mounted on a modified M48A5 tank chassis, has two 40 mm guns with search and track radar, a fire control center with laser rangefinder and a digital computer. (Ford Aerospace via M G Burns)

The Royal Air Force's BAe Nimrod AEW is based upon the DH Comet airliner, and the trial conversion is shown here. The massive nose and tail radomes house the radar scanners, while the operators are accommodated in the fuselage. (British Aerospace)

The US Air Force's Boeing E-3A Sentinel AWACS aircraft is based on the Boeing 707 airliner. It is capable of monitoring a vast area and controlling warplanes within it.

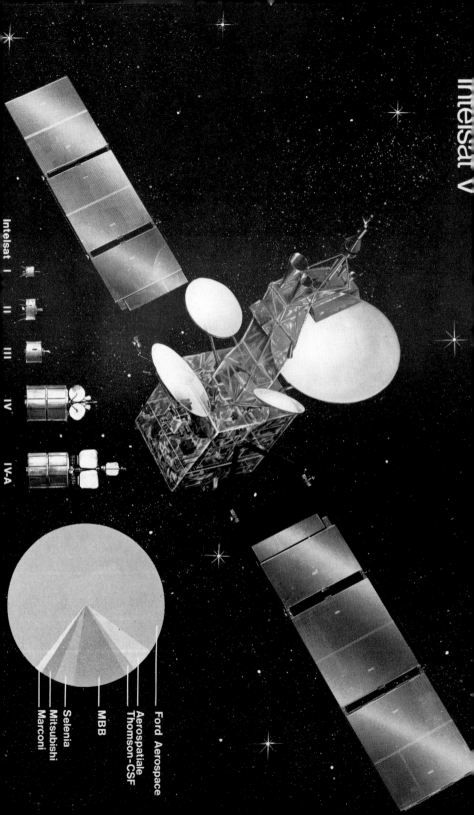

Intelsat V

Intelsat I

II

III

IV

IV-A

Marconi
Mitsubishi
Selenia

MBB

Aerospatiale
Thomson-CSF

Ford Aerospace

Two Soviet warships in port, cluttered with electronic warfare scanners, sensors and aerials. Modern warships depend upon comprehensive EW equipment for their survival against a multiplicity of threats, most of which can develop very rapidly, while their offensive and defensive weapons systems are based upon electronics and can only be deployed effectively through the use of EW systems.

A Soviet guided-missile cruiser in port, weighed down with electronic warfare equipment. Able to launch its missiles at stand-off range, the cruiser requires sophisticated target detection systems.

Ford Aerospace is the prime contractor on the Intelsat V new generation of world-encircling communications satellite systems, but, on a project of this magnitude, international co-operation is normal. (Messerschmitt-Bölkow-Blohm GmbH via M G Burns)

An artist's impression of the first reusable satellite, the MBB private-venture SPAS-01, being lifted out of the US Space Shuttle for a free flight phase. (Messerschmitt-Bölkow-Blohm via M G Burns)

would allow their pilots to penetrate that ring of fire which the Arabs had set up with some chance of successfully accomplishing their mission and surviving. Most of these tactics involved attacking the enemy missile system directly. A very risky but effective attack manoeuvre developed for a single-aircraft attack on a SAM-6 system took advantage of the SAM-6's poor low elevation capability and the slow speed at which it could be elevated. The aircraft flew towards the launching vehicle at an extremely low altitude to avoid detection by the system's anti-aircraft radar, hiding in the false echoes produced by ground reflection ('ground clutter'). Once he had passed the target, the pilot had to pull up sharply into an almost vertical climb, and then immediately dive at the target, launching his missiles or bombs at the right moment. During this dive and his subsequent escape, the pilot, still at an extremely low altitude, had to launch first chaff to deal with any SAM-6 missiles which might be launched against his aircraft and then carry out further evasive manoeuvres to avoid SAM-7 IR-guided missiles. The simplest of such manoeuvres was to launch flares and then turn towards the missile so that the jet tailpipe, the hottest part of the aircraft, would be pointing away from the missile.

Even more complicated techniques were used. One involved two aircraft flying side by side which, as soon as they realised that an IR-guided missile had been launched (or were informed of the fact via radio by helicopters patrolling the area), carried out a manoeuvre which involved one of them intersecting its own previous flight path, thus creating a zone of intense heat which, being IR energy, attracted the SAM-7 missile.

Another very effective tactic exploited the limited rate and range tracking ability of the SAM-6 system. A Phantom and a Skyhawk would approach at a high altitude, one behind the other: the first aircraft, the Phantom, would launch a large quantity of IR-flares and chaff to jam enemy radars and guidance-systems thus enabling the Skyhawk to dive on the target and release its bombs or missiles with a good chance of success and survival.

All these tactics relied upon drastic, almost desperate, manoeuvres which the missile's guidance-system was unable to follow; such aerobatics demanded excellent reflexes and coordination of the pilots.

Later, the aircraft were equipped with new RWR in pods capable of intercepting the very high frequency electromagnetic emissions emanating from the SAM-6 batteries and the *Shilka* AAA batteries' fire-control radar.

With these new systems, the Israelis not only reduced their aircraft

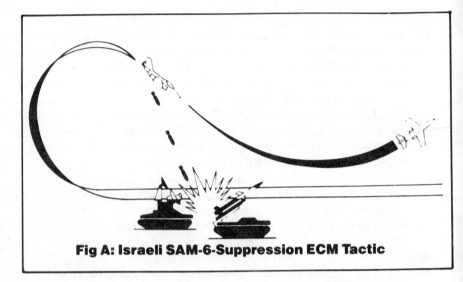

Fig A: Israeli SAM-6-Suppression ECM Tactic

A. An ECM tactic used by the Israelis to attack SAM-6 batteries and avoid their missiles. Exploiting the weak points of the SAM-6 system, mainly the battery's inability to respond well at very low elevation, the aircraft approached at an extremely low altitude, concealed by the confusing radar echoes reflected from the ground (ground-clutter). After passing the battery, the aircraft pulled up vertically and then nose dived onto the target, launching anti-tank missiles or bombs. On completion of the attack, the aircraft, still flying very low, launched chaff and IR flares to deceive any other SAMs which might be launched at it.

loss-rate considerably but also managed to destroy forty out of a total of sixty missile systems. Having regained air superiority, previously held by the Egyptian and Syrian anti-aircraft missile systems, the Israeli air force was once again able to provide tactical air support for their ground forces, not only defensively to block the advancing Arab forces but also offensively during the famous Operation 'Gazelle' in which the Israelis crossed the Suez Canal and penetrated deep into Egyptian territory.

At the close of hostilities, the final toll was 110 Israeli aircraft lost, a high figure considering the size of the Israeli air force; most of these had been shot down by the new weapons systems which had taken the Israelis by surprise, finding them without effective electronic and infrared countermeasures.

The results of Arab-Israeli sea warfare were, however, quite different. We have already seen how, during the Six-Day War, the

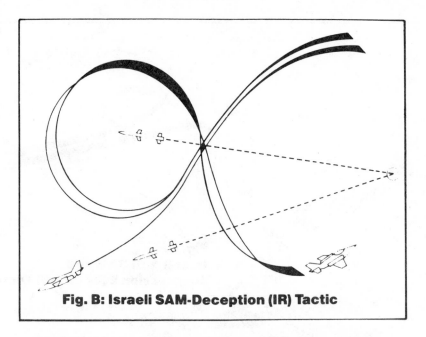

Fig. B: Israeli SAM-Deception (IR) Tactic

B. The Israeli air force used IR flares to avoid IR-homing missiles, employing various techniques. One technique involved two aircraft, one of which would suddenly veer upwards and then nose dive, crossing its own flight path and thus creating an area of great heat. Meanwhile, the other aircraft flew on, launching IR flares. It was also found that sudden drastic changes in its course constituted an effective means of defence, tending to deceive the missiles.

Israeli destroyer *Eilat* being equipped with neither RWR nor ESM (Electronic Support Measures, without which ECMs could not be carried out) nor chaff, nor other jamming equipment, had been sunk by Soviet-built Egyptian fast patrol boats launching a salvo of Soviet-built *Styx* missiles against the unfortunate ship without even leaving port. After this disaster, it was decided to update and improve the Israeli navy. The first step was to start construction of a new class of warship, the 'Reshef' class of Fast Attack Craft. These displaced 410 tons and were armed with launchers for locally made Gabriel missiles.

On the other side, the Egyptian and Syrian Navies had a large number of Soviet-built 'Komar' and 'OSA' class vessels all equipped with *Styx* missiles which, until then, had never once failed to hit their targets. They had proved their worth in the 1971 Indo-Pakistan war when, between 4 and 8 December, numerous Pakistani warships in the

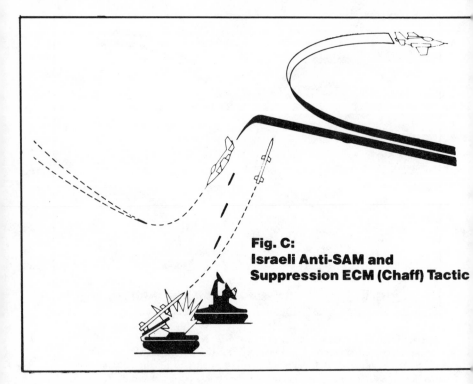

**Fig. C:
Israeli Anti-SAM and
Suppression ECM (Chaff) Tactic**

C. Another Israeli two-aircraft tactic for avoiding enemy missiles. The first aircraft turns back, launching chaff to deceive enemy radars, while the other carries out a dive-bombing attack on the missile battery.

area of Karachi, as well as three merchant ships anchored in the port itself, had been sunk by Styx missiles launched from Indian 'Komar' and 'OSA' class boats.

The Israeli Gabriel missile was more accurate than the *Styx* but its range was decidely inferior, by a ratio of 2:5. In practical terms, this meant that an Israeli 'Reshef' or 'Saar' class boat carrying Gabriel missiles would have to penetrate a danger zone of 20–30 kms, in which it would be within range of enemy *Styx* missiles, before it could launch its own. It was, therefore, imperative to find a workable tactic for combatting enemy naval squadrons armed with *Styx* missiles. The search for such a tactic became the prime concern of the Israeli navy.

Experience had shown that the problem could not be solved by traditional defence systems which had proved themselves to be impotent in the face of these anti-ship missiles. The Israelis soon recognised that something completely new was required and that the solution lay in the field of ECM.

Consequently, they equipped all their missile-launching boats with electronic jammers and deceivers and covered their ships with material which partially absorbed, rather than reflected, energy from any radar waves which might strike it. Such radar absorbent materials (RAM), called 'microwave absorbants', were able to irreversibly transform energy from electromagnetic waves into another type of energy, in this case heat, which could be easily dispersed into the air and water. It was also decided that the best manoeuvre to adopt, during attack, was to face the enemy bow-first so as to present the enemy radar with the smallest possible reflective surface.

When the Yom Kippur war broke out, the small but well-armed navies of the Middle East were well-prepared for that series of naval conflicts which remain unique and of great importance in modern naval history.

On that first night of the war in October 1973, the Israeli naval command, fearing that the Syrians might launch a naval attack on the port of Haifa, ordered five fast attack missile-boats—*Reshef*, *Mivtach*, *Hanit*, *Gaash* and *Miznag*—to sail north and search for enemy units.

The Syrians, for their part, were also worried about their own lack of coastal defences and sent out three 'OSA' and 'Komar' class missile-boats, as well as various other vessels, on surveillance and patrol missions.

The Israeli formation skirted the Lebanese coastline to reach Syrian waters where, at 22.28, they sighted a Syrian torpedo-boat on patrol off the coast near Latakia. The Syrian boat tried to take refuge in a nearby port but was caught and sunk by gunfire from the five Israeli ships.

The Israeli formation then turned eastwards, splitting into two columns to commence a sweep in the direction of Latakia (see page 198). During this phase of the mission, *Reshef* sighted a Syrian minesweeper which it promptly sank with one of its missiles. But the minesweeper was probably a bait to attract the Israelis towards the three Syrian missile-boats which were preparing to attack the Israeli formation.

The ESM equipment on board the Israeli vessels gave the alert and analysis of the radar emissions intercepted furnished data regarding the attacking vessels' type and armament. Both the Syrian and Israeli formations immediately manoeuvred towards favourable firing po-

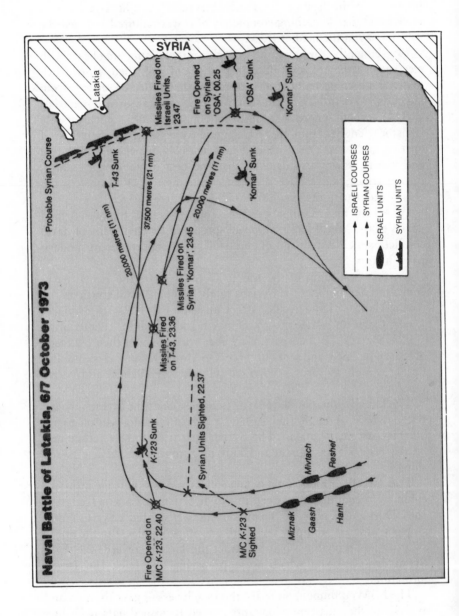

Naval Battle of Latakia, 6/7 October 1973

sitions. They were now 25 miles from each other, but the distance closed rapidly as they raced towards each other at full speed.

At this point the Syrians enjoyed the advantage since, unlike the Israelis with their shorter range Gabriel missiles, they were now within launch range for their *Styx* missiles. They launched their first Salvo from a distance of 37,500 metres and the Israelis immediately activated their Deception Jammers to send the *Styx* missiles off course and launched quantities of chaff to further distract them. The Israelis fired both long and short range chaff in their prepared plan to create maximum confusion for the *Styx* seeker heads.

There was great tension among the crews of both the Israeli and Syrian units who were well aware that their fate now depended on the electronic equipment they had on board—the missiles on the Syrian side and the deception jammers and chaff on the Israeli side. It was the first battle in naval history between two missile-launching formations and there was no telling what might happen! This was not a classic naval battle in which gunfire was directed and corrected by men; the result of this encounter depended on electronic equipment, masterpieces of technology which could do incredible things but which, nevertheless, each had shortcomings. Missiles need radar to lock onto and track the target and radar is vulnerable to ECMs.

The sinking of the *Eilat* had taught the Israelis the great importance of electronic warfare and they had learnt their lessons well. As soon as the ECM equipment on board their ships was put into action, the Syrian *Styx* missiles immediately deviated away from their real targets towards non-existent targets and, after wild and uncontrolled manoeuvres, crashed harmlessly into the sea.

Having avoided the first missile attack, the Israeli ships continued at full speed ahead, in two lines, until they came within launch range of their own Gabriel missiles. They opened fire at 23.36 and the Syrian boats, lacking electronic warfare equipment of the kind used by the Israelis, suffered heavy damage. One 'Komar' craft and one, 'OSA' craft sank shortly thereafter, while the other 'Komar' craft drifted onto a sand-bank where it was then destroyed by gunfire from two of the Israeli ships.

The following evening the Israeli navy took part in another naval battle in even more dramatic circumstances, this time against the Egyptians. The Israelis had discovered, by intercepting enemy communications, that an Egyptian naval formation was going to move out of Alexandria that night and sail to the naval base at Port Said, nearer to the front. The Israeli high command immediately sent out their missile-boats *Reshef, Keshet, Eilat, Mifgav, Herev* and *Soufa* to intercept and destroy the enemy naval formation.

The Israeli ships sailed towards the Egyptian coast, maintaining strict radio and radar silence; only passive electronic warfare equipment was in operation, that is, those which emit no electromagnetic energy (RWR and all ESM receivers).

The Egyptian formation, consisting of four 'OSA' class attack craft armed with *Styx* missiles, left Alexandria just after sunset and headed for Port Said. At about 21.00, one of the Egyptian ships switched on its radar for a few seconds to check the route and to find out whether there were any enemy ships nearby. This electromagnetic 'indiscretion' was promptly picked up by the Israelis, informing them of the presence and location of the Egyptian formation.

Navigating in total darkness, the two naval formations drew closer. At 23.00, the Egyptians picked up the six enemy units on their radar scopes at a distance of approximately 26 miles. As soon as they were within firing range, 24 miles, the 'OSA' gun-boats launched a salvo of twelve Styx missiles. However, the Israeli ships' ECM devices—noise and deception jammers and chaff-launchers—sent all twelve missiles off course and they ended up in the sea.

The Israeli units sailed on at full speed towards the enemy and, after twenty minutes, were near enough to launch their own missiles. The Egyptians, having no ECM equipment on board their 'OSA' boats, were powerless to counter the Gabriel missiles launched against them and three of their boats were hit and sunk. The fourth unit was badly damaged and drifted onto a sandbank near Baltim.

The importance of the role played by ECM in these battles needs no comment! The opposing naval formations never came within visual range of each other; everything was done electronically and, in each engagement, the side with the more effective ECMs emerged victorious.

In the naval battles of Latakia and Damietta-Baltim, none of the fifty-two *Styx* missiles launched against Israeli units hit their target, a fact which speaks for itself. These results were due to the planning and efficient use of EW equipment by the Israeli navy, and they brought to an end the threat of the *Styx* missile for the navies of the Western Powers.

While these naval battles did not greatly affect the outcome of the Yom Kippur war, they certainly marked a turning-point in the history of naval warfare.

The participation of Russia and the USA, although they did not

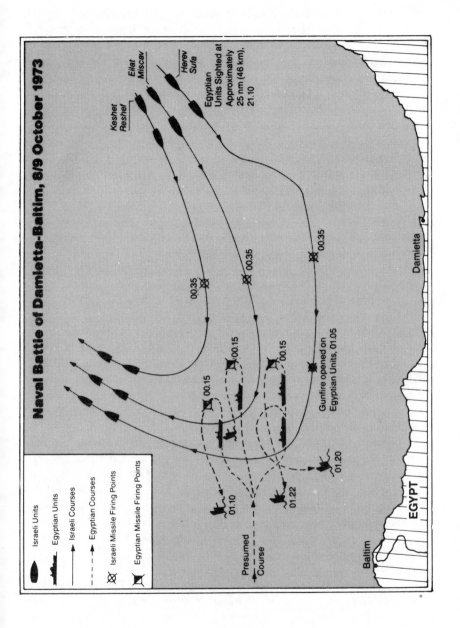

Naval Battle of Damietta-Baltim, 8/9 October 1973

Israeli Units	
Egyptian Units	
Israeli Courses	
Egyptian Courses	
Israeli Missile Firing Points	
Egyptian Missile Firing Points	

officially intervene in the war, at any point was, nevertheless, of crucial importance. The two 'Superpowers' did much more than merely provide arms, electronic systems, logistic support and so on; they used the Middle East like a huge 'missile-range' where they could try out their latest arms and equipment.

It is a proven fact that the Americans used Israeli aircraft to evaluate, in real tactical situations, AGM-65 maverick missiles, which are guided by laser-beams and rarely miss their targets. Similarly, a new version of the AGM-45 Shrike air-to-surface missile, which guides itself onto the enemy radar by 'riding' the radar's own electromagnetic emissions or similarly 'rides' enemy jamming signals (Home-on-Jam) to source, were also tested in real combat situations.

The Russians tried out their new air-to-surface missile, AS-5 *Kelt*, which has a range of over 200 miles, on the very first day of the war. An Egyptian-piloted Tupolev Tu-16, flying over the Mediterranean, launched one such missile in the direction of Tel Aviv. By sheer chance, the missile was sighted by an Israeli Phantom which intercepted it and shot it down.

The Soviets also used the Middle East war to test the effectiveness of their wire-guided anti-tank missiles, *Snapper* and *Sagger*, and the latest version of their *Frog* surface-to-surface missiles (*Frog* 7), which were used on the Syrian front. Similarly, the US anti-tank missile, TOW (Tube-launched, Optical-tracking, Wire-guided), was tried out by the Israelis. Tow is a system consisting of a tube-launcher and an optical-tracking device attached to a tripod. The missile is wire-guided and controlled by a computer which automatically sends route-signals to the missile in flight.

New aircraft were also tried out in the Yom Kippur war. Special reconnaissance missions were carried out by the US Lockheed SR-71 Blackbird spy-plane, which has a speed of over Mach 3 and a ceiling of nearly 100,000 feet, and also by the Soviet MiG-25 *Foxbat-B* which has a speed of Mach 3.2 and a ceiling of approximately 80,000 feet. Several supersonic Sukhoi Su-25 *Flagon-A* and Su-20 *Fitter-C* fighters were also reported to have been sighted over Israeli territory, as was the most recent version of the French Dassault Mirage over Arab territory.

Pilots from neutral countries were also involved in the war for various reasons, such as training, testing new equipment and acquiring first-hand experience of the latest air tactics. Interception of flight communications revealed the presence of Pakistani, Cuban and Libyan pilots on the Arab and South African pilots on the Israeli side.

A lot of hasty conclusions have been drawn from the experiences of the Yom Kippur war. For example, it has been said that the advent of missiles marked the end of tanks and aircraft, but it is outside the scope of this book to evaluate the validity of such an assertion. However, the experiences of this war can teach valuable lessons about electronic warfare.

One of the most important teachings of the Yom Kippur war regards the extremely serious consequences which can derive from an inadequately functioning intelligence service. The Israeli intelligence services were accused, not unjustly, of failing to provide the government with sufficient warning of the imminent Egyptian-Syrian attack, a failure which threatened the very existence of the Israeli nation. A second serious shortcoming was that the Israeli armed forces found themselves without adequate ECMs to counter the enemy's new, sophisticated electronic weapons systems, and this resulted directly in very severe losses of both men and equipment.

All this could have been avoided if the electronic sector of the intelligence service, called SIGINT, had been more efficient; this is surely the duty of any State concerned about its own security and survival. It is impossible to know precisely whether the shortcomings of the Israeli SIGINT service lay in data-gathering or in evaluation and analysis. Nevertheless, it is certain that, if the Israelis had been more thorough in their interception and deciphering of Arab communications and analysis of radar emissions in peace-time, they would not have suffered those terrible twin surprises—the attacks themselves, and the new-generation weapons systems deployed by the Arab forces.

Egypt, on the other hand, certainly did not neglect military intelligence but, rather, had made excellent use of it before the outbreak of the war. After the bitter experience of the 1967 war, they had no intention of being caught by surprise again. With Soviet help, they completely modernised their intelligence service, first, by acquiring all kinds of up-to-date equipment for electronic espionage: highly sensitive radio receivers, radar surveillance receivers, tape-recorders, direction-finders and so on.

During the war, the Israelis managed to capture, among other things, Egyptian maps showing, in great detail, their defence installations, planned operations along the Canal, the code-names of bases in the Sinai and so on. They also had the good fortune to capture a number of complete SAM-6 systems, SAM-7 missiles and ZSU-23-4 batteries which provided them with precious information about their respective radars and the technological level reached by the Soviets in the field of electronic warfare.

The Yom Kippur war is an excellent example of limited, as opposed to general, warfare. It was a war with a limited objective, limited time and limited space, sponsored by two Superpowers who wanted to try out their latest weapons. The presence of so many electronic devices controlling the various weapons systems made it extremely difficult to keep the situation in check; this was further aggravated by communications jamming, particularly by the Egyptians. In fact, lack of air control led to several instances of both sides shooting down their own aircraft. This last problem should be kept in mind by those responsible for planning future defences and weapons systems, since the kind of incidents which took place in the air could also happen in ground or naval warfare, with more serious consequences.

To sum up it is essential for all armed forces to be equipped with a complete range of EW equipment, even in peace-time, and to have an efficiently-run intelligence service, with up-to-date equipment for electronic espionage (SIGINT), able to stay constantly abeam of the technological progress of potentially hostile nations.

18

Infrared

The various ECMs introduced by the Americans in the last few years of the long war in Vietnam almost managed to neutralise the effectiveness of radar as a means of detection and guidance. The *Fansong* radar used to guide SAM-2 missiles was often completely jammed or deceived by ECMs, while the Soviet air-to-air missiles arming MiGs were impotent against US aircraft equipped with 'smart' or deception jammers. In consequence new missile-guidance systems exploiting infrared (IR) energy were researched and developed.

The use of this form of energy was not new. It had been discovered by chance, in the year 1800 by the British astronomer Sir William Herschel, who was already famous for his discovery of the planet Uranus. He was experimenting with various coloured glass filters to protect his eyes from the sun's rays, which caused him considerable discomfort during his astronomical observations. During these experiments, he noticed that heat reduction was not equivalent to light reduction. Therefore, he devised an experiment in which the solar spectrum was projected onto a screen by passing light through a glass prism. When he passed a thermometer over each of the projected colours, he noted that the temperature increased as the thermometer passed from blue to red. He further noted, with some surprise, that, after passing through red and into the 'empty' zone, the thermometer continued to show an increase in heat; this area has since become known as the infrared spectrum. He had, in effect, discovered that the solar spectrum contained rays other than those which could be seen by the naked eye and he therefore called these 'invisible rays'. Herschel did not fully appreciate the importance of his discovery, however, and many years passed before it was followed up by further experiments; this time lapse can also be attributed to the lack of instruments for measuring heat, apart from the common thermometer.

During World War One, considerable progress was made in developing practical applications of infrared rays. Both sides were quick to realise the military importance of infrared radiation as a means of seeing in the dark without being seen, of detecting targets by their heat emissions and of conducting secure communications which were very difficult to intercept. A signalling system using infrared pulses with a range of 2 miles, and a night vision device capable of detecting an aircraft at an altitude of 5000 feet, or a person at a distance of 900 feet were developed in those years, although only to the experimental stage.

Research in the infrared field really gained momentum during World War Two. It is interesting to note that this momentum was set off by an error of judgement by the Germans during the Battle of the Atlantic between Allied convoys and German submarines. Allied anti-submarine forces stopped using search radar operating in the L-band because these emissions were so easily intercepted by German submarines who, thus alerted, crash-dived and escaped. They introduced radars operating on a higher frequency, in the X-band, and this had, of course, immediately resulted in an increase in German submarine losses which the Germans could not understand. The German secret service was called in to find an explanation and they erroneously concluded that the Allies were using infrared ray detectors.

This mistaken conclusion caused the Germans to waste a lot of time and, no doubt, contributed to the final defeat of their sumbarines in the Battle of the Atlantic. On the other hand, German efforts devoted to the study of infrared radiation led to important advances being made in this field. Many people in Germany still remember the fear and amazement they experienced when huge armoured vehicles roared past them at night, with no lights! These vehicles were transporting the infamous V-1 flying-bombs and their launchers to the French coast of the English Channel. To avoid being detected by enemy aircraft, they were using a device comprising an infrared emitter and an image converter, to enable the drivers to see in the dark. This phenomenon took place towards the end of the war when Germany was suffering constant air raids.

The Germans also used infrared rays in ship-to-ship, ship-to-shore, and, on the Libyan Front, tank-to-tank signal communication systems. However, during the Battle of El Alamein in 1942, one of these systems was captured by the British and, thereafter, the Allies also began research into the use of infrared radiation for military purposes.

The Americans used an infrared ray device for aiming rifles in the dark: it afforded sufficient accuracy to hit a man at a range of approximately 80 yards. This weapon, called a Sniperscope, was first used by American soldiers during beach landings in the Pacific, causing great terror among the Japanese soldiers.

In Italy, infrared ray devices were first evaluated, experimentally, by the navy in 1941–42 for the purpose of determining the distance at which a target could be detected in conditions of darkness or fog. The device used was a receiver consisting of a parabolic mirror with a diameter of 50 cms incorporating a thermo-electric detector cell. Night experiments demonstrated that a person could be observed at a distance of about 100 yards and a vehicle with its engine running at about 500 yards; the cruiser *Taranto* was sighted at a distance of 5000 metres even though she was not using full engine power at the time.

Research on infrared systems continued after the war, its value as a means of detection which could not itself be detected now being fully appreciated. Continual progress in this field has brought about a long series of inventions for military use.

In the field of aeronautics, an IR tracking device able to signal the elevation and azimuth (bearing) of any heat-emitting target, in the air, on the ground and on or under water, was developed; it could also be used as an aid for instrument landing systems (ILS) for aircraft and in hydrographic surveys along coasts. Another important invention was Forward Looking Infra Red (FLIR). This equipment enables a pilot flying in cloud or total darkness to 'see' all objects on the ground or below the clouds having a different radiometric temperature from their immediate environment. IR systems have proved to be extremely useful in the field of missiles and strategic surveillance; installed on satellites, they give immediate warning if an Intercontinental Ballistic Missile is launched from any point on earth. Devices able to detect the presence of noxious or poisonous gases in the atmosphere have been developed as surveillance aids. IR sensors have been added to radar antennae to improve their performance, especially when radar silence has to be observed.

Military demand for IR devices led to accelerated technological progress in this field and the development of ever more sophisticated devices, such as power sensors, radiometers and other IR measuring instruments.

The applications of IR in the purely scientific field, in industry and in medicine are too numerous to mention here. IR systems are used for widely different purposes from testing the asphalt surface of roads to early diagnosis of tumours and many other diseases, particularly vascular illnesses, from infrared ovens for cooking food to systems in chicken incubators, from spray painting of cars to measuring the temperature of the stars. One of its best known uses is certainly in photography, the first experiments with IR being made in the 1930s, since which time there has been a whole series of innovations in this field. For example, good photographs can be taken, using IR techniques from a distance of over 1000 kms on a day when visibility is only 10 kms; this is particularly useful in the field of geodesy. IR photography is also useful in checking the health of plants by the colour of their leaves which shows up clearly in IR photographs, enabling one to distinguish sick plants or trees from healthy ones.

In geology, IR photographs of the stratigraphic layers of the earth reveal its geological age since the presence of fossils in the rock shows up clearly, because fossils and the earth covering them have different radiometric temperatures. Analogous techniques can be used to detect underground operational centres and ammunition stores as well as archeological objects and the remains of buried cities. IR techniques are extremely useful in detecting counterfeit letters and documents since some types of ink are revealed by IR rays while others are not. Applications of IR in the field of communications are very interesting; research is aimed at developing systems whereby signals are transmitted by electromagnetic waves in the infrared spectrum via optical fibres. These applications are of particular interest to all sectors of telecommunications such as the telephone, videophone, cable television and data transmission.

In order to examine further the applications of infrared energy, it will be useful first to recall certain notions which belong to the field of physics.

It is known that the retina of the human eye is sensitive to only a small sector, i.e. the visible sector, of the electromagnetic spectrum; furthermore, the eye's sensitivity is not constant but varies according to the chromatic scale of light. For example, the stimulating effect of yellow light is almost 100,000 times greater than that of red light which is one of the weakest colours in this respect. The wavelength of yellow light is approximately 0.0005 millimetres; with both longer and shorter wavelengths, the eye's sensitivity drops gradually. At the longer wavelength end of the spectrum, the eye is still able to pick up wavelengths of about 0.0008 millimetres but, beyond this point, darkness reigns since the stimulus of such radiation is too weak to produce a response in our visual organs. Waves longer than 0.0008 millimetres are in the infrared region and, if they are of sufficient intensity, are perceived by us as heat. The main factor which distinguishes infrared from light radiation is, therefore, the wavelength. The infrared region extends from the point where the red end of the visible spectrum ends to the microwave band used for high

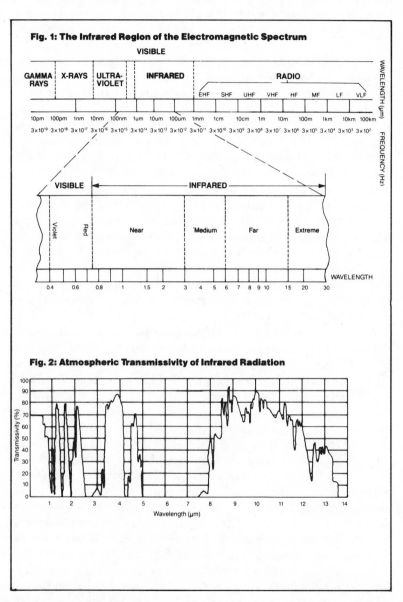

Fig. 1: The Infrared Region of the Electromagnetic Spectrum

Fig. 2: Atmospheric Transmissivity of Infrared Radiation

Infrared rays can easily pass through bands between 3.0 and 4.0 µm but are greatly attenuated by the band between 5.0 and 8.0.µm. In other words, the atmosphere is 'transparent' to infrared radiation only for certain wavelengths.

resolution radars (EHF). The infrared region is itself divided into four parts: near, medium, far and extreme. The main factors involved in an infrared system are the sources, transmission of IR energy and detectors or sensors.

IR energy is spontaneously emitted by all bodies that have a temperature above absolute zero (− 273 degrees Celsius). The process is set off by atomic oscillations in the molecules composing the bodies and is therefore closely related to their temperature.

A prototype infrared detector is to be found in nature, in the animal kingdom. The last of the snake families to evolve, the highly poisonous pit viper family, found in North and Central America and South-East Asia in particular, has two small dimples between the eyes and nostrils containing two perfect IR sensors which enable it to detect and locate all things which are either hotter or colder than the immediate environment. These sensors are extremely sensitive and can detect minute variations in temperature. They are composed of a membrane, full of special nerve fibres which react to heat, stretched over a small air-filled cavity. The snake, which usually hides in holes dug in the ground, is thus able, in total darkness, to detect the presence of a frog, mouse or any other unfortunate creature which has come within its sphere of action, and kill it.

Being a form of electromagnetic energy, IR radiation can be absorbed and transformed into heat or processed in such a way as to make it observable; for example, it can be transformed into electric current or projected onto photographic film susceptible to infrared rays. From a military point of view, IR energy detectors demonstrated their practical usefulness towards the end of World War Two when they were first used to detect aircraft and follow their route. IR devices used in World War Two were nearly all of the active type, a beam of infrared radiation being focussed on the target. However, the Germans did also experiment with a completely passive system which did not emit IR energy itself but, instead, detected the IR energy emitted by the target itself—like the rattlesnake. This system was designed to detect aircraft at a distance of 12 kms but was never actually put into operation, probably because IR technology was not yet sufficiently advanced to permit actual production of such a system.

After the war the major world powers continued their research on infrared rays, concentrating on passive systems for the guidance of weapons. These systems had the advantages of not revealing the presence of the guided weapon, of affording a high degree of accuracy and, above all, of being immune to ECM. In 1950, this research led to the development of the first passive IR missile-guidance systems. Missiles guided by such systems are the American AIM-9 Sidewinder, the first, and AIM-4 Falcon, the British Firestreak and the French Matra R.550 Magic.

The best known of these missiles is the AIM-9 Sidewinder which gave proof of its great accuracy in the very first tests carried out. The radio-guided target drones used in the tests were systematically destroyed by the missile which homed right into the exhaust nozzle of the targets jet engine: to reduce the replacement costs of the expensive drones used in these trials, strong IR sources were fixed to the wings, which were much cheaper to repair. Further confirmation of the accuracy of the Sidewinder was provided by a tragic incident which took place during a training exercise in the United States in 1961. A B-52 Stratofortress was shot down by a Sidewinder missile accidentally launched from a USAF F-100 fighter-bomber; the missile homed onto the exhaust of one of the jet engines causing an explosion which tore off a wing and sent the bomber hurtling to the ground. Most of the crew was killed.

Several years later, a Sidewinder missile was involved in an incredible espionage affair, many aspects of which are still unclear. An enterprising Soviet agent somehow managed to steal an entire Sidewinder missile with its IR guidance seeker head from an airbase in West Germany and smuggled it to Moscow. He travelled half way across Germany with the missile rolled up in a carpet in his car and then sent it across the border by train as unaccompanied baggage 'of no commercial value'! The Russians, soon thereafter, produced an IR-guided missile of their own, the AA-2 *Atoll*, which was almost identical to the American Sidewinder.

Atoll missiles carried by MiGs and Sidewinders carried by various US fighters were the main air-to-air weapons used during the Vietnam War. A Sidewinder was responsible for shooting down the first North Vietnamese MiG-21 in 1966.

In 1973, a few months before war broke out in the Middle East, about a dozen Syrian aircraft were shot down by Israeli Shafrir IR-guided missiles, which is based on the Sidewinder. In October 1973, the Israelis themselves suffered heavy losses on the Egyptian front caused by the Soviet-built SAM-7 *Strela* IR-guided missiles. The *Strela* missiles were equipped with filters which eliminated, to a certain extent, one of the main defects of IR guiding systems: the high percentage of false alarms resulting from distraction by other heat sources. They were very easy to handle and could be carried by a single soldier on his shoulder. They proved to be a deadly weapon against Israeli aircraft which were forced to fly low to avoid search radars and SAM-6 missile-guiding radars. Fortunately for the Israelis, the explosive power of the Strela was limited due to the small size of its warhead; otherwise, there would have been a real massacre of Israeli aircraft.

However, IR-guided missiles had several shortcomings. The most serious defect of the Sidewinder was that, instead of homing in on the target, it would often head for stronger sources of heat such as the sun itself, sunlight reflected by clouds, heat-producing bodies on the ground or even, in some cases, nearby friendly aircraft. There was also the serious limitation of having to attack the enemy aircraft from virtually dead astern in order to present the missile with the hottest area of the target, the jet pipe.

These serious shortcomings arose from the fact that the IR-sensors fitted to the missile seeker head were not sensitive to the right IR wavelengths. For example, they had to attack enemy aircraft from behind because the lead-sulphur sensors used on the first Sidewinder missiles only reacted to wavelengths corresponding to the hot metal of exhaust nozzles of jet engines. An IR sensor which would react to the whole jet stream coming from the engine was needed so that the missile could be launched regardless of the position or direction of the enemy aircraft. In technical terms, it was necessary to devise an IR sensor which would react not to wavelengths of 2.5 μm (corresponding to sun-rays reflected by clouds or emissions from the incandescent metal of the jet exhaust nozzle) but rather to wavelengths of 5 μm, corresponding to exhaust gases. This was done by freezing the detector itself to temperatures referred to as cryogenic (from the Greek word *Krios*: intense cold), such temperatures being much lower than those obtained by normal freezing units.

Most of the original defects of IR-guided missiles have now been overcome by various means, such as the use of filters, and they are now a vital weapon in the arsenals of many countries. Combined guidance systems are now often used in which radar is used to measure distance while IR is used for direction finding or as a secondary system in case the radar is neutralised by ECMs. IR is also used to distinguish a 'hot' target (for example, a ship or an aircraft) from one which does not emit heat (for example, chaff).

Looking back at the past few decades, we can note that every time a new weapons system entered the picture, parallel countermeasures were devised to neutralise or reduce its effectiveness. This happened first with radar systems and is now happening with IR systems. Information regarding IR countermeasures (IRCM) is difficult to come by however, as such developments are kept top secret. Nevertheless, it is certain that many nations are devoting a considerable intellectual and financial effort to developing countermeasures to IR-guided arms systems.

IR Warning Systems, which have exactly the same function as RWRs, are already in existence. When installed on aircraft, they warn the pilot of the approach of a missile: the IR sensor either picks up the heat (IR energy) emitted by the missile on launch or during its thrust phase, or picks up the friction produced as the missile passes through the atmosphere. Such early warning enables the pilot to make an appropriate evasive manoeuvre, launch flares or activate IR devices to interfere with the missiles flight path, should it be IR-guided. Such devices for jamming or deceiving IR seekers are based on new concepts, such as laser beams which can damage or even burn-up the IR sensor of an enemy weapons system. Another IR CM involves burning combustible material in combustion chambers to heat a special membrane which radiates specifically modulated IR energy. Other systems use propane gas, burnt in special containers, or arc lamps to produce IR energy which interferes with enemy missile-guidance systems.

During the Yom Kippur war, IR deception was used with considerable success. Pyrotechnic devices, such as IR flares and decoys, which emitted greater heat than the targets they were defending, but with the same IR characteristics, were deployed or dispensed to lure IR-guided missiles off-course. A whole new field of electronic warfare had opened up.

19

Laser, Television and 'Smart Bombs'

Lasers

At the beginning of the twentieth century, a Croatian physicist named Nikola Tesla,[1] who had emigrated to the United States, invented a transformer (named after him) which had an extremely high ratio of transformation and was capable of producing extremely high tension in the region of hundreds of thousands of volts. Military authorities all over the world showed great interest in this discovery since, according to Tesla he had invented a kind of 'death ray' capable of causing the disintegration of whole formations of aircraft at a distance of 300-400 km.

At first, it was thought that the much longed-for 'absolute weapon' which could win all wars had finally been invented. But the initial enthusiasm soon died down when the brilliant but eccentric physicist failed to furnish details of his revolutionary weapon. Nevertheless, the military commanders of major world powers were unwilling to let go of the idea of a 'death ray' and waited year after year for their dream to come true.

On 26 February 1935, leading figures from the British War Ministry were invited to one of the main military radio stations near London to watch the physicist Robert Watson-Watt give a demonstration of radar. This event caused great excitement since Her Britannic Majesty's General Staff had placed a very specific request; they explicitly asked whether radar could produce 'death rays' which would guarantee the supremacy of British armed forces over all potential enemies. Although they had before them a great invention which was to revolutionise traditional warfare, those present were somewhat disappointed when the demonstration fell short of their expectations.

Many years later, in 1960, the research laboratories of the American Hughes company produced the first laser generator machine, developed by the physicist Theodore Maiman. Again there was talk of a 'death ray' and many journalists really went to town on the subject.

However, one of the first uses of laser was in the field of medicine—in microsurgery, where laser beams were used to perform extremely delicate operations, such as in brain surgery, eye surgery for repairing detached retinae, in the treatment of certain forms of cancer by destroying the malignant tissue, in stomatology and endoscopy. Laser has also proved to be of great importance in science and technology— in the fields of spectroscopy, microanalysis, high-speed photography, microphotography, microwelding and precision engraving, to give just a few examples.

Of course, laser also has many important applications in the military field, which exploit characteristics of the laser beam different from those exploited in civilian fields. One important military application of laser is in very high-precision weapons guidance: guidance of 'smart' bombs, or Laser Guided Bombs, such as the US Texas Instruments Paveway LGB, and missiles, such as, for example, the US-Hughes AGM-65 Maverick. These are fitted with trackers which home on to radiations emitted by a target which is being illuminated by another laser beam called a laser-designator. The tactic generally used to launch a smart bomb is the following: two aircraft are used, one equipped with a laser-designator, which illuminates the target with a coded laser beam, while the other releases a pre-programmed bomb which homes on to the laser energy reflected by the 'illuminated' target, hitting it with near perfect precision. Alternatively, the laser

Illustration of a two aircraft attack. One aircraft is equipped with a laser designator to illuminate the target with a coded emission; the other carries laser-guided bombs (LGB) which, upon release, 'ride' the laser reflection and, thus, are guided onto the target. The use of an individual code by each pair of aircraft permits several pairs to operate in one zone simultaneously.

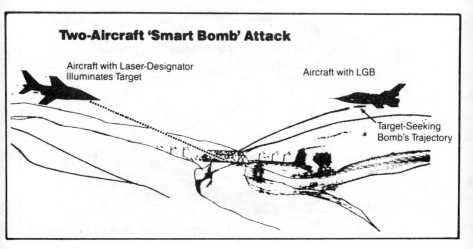

Two-Aircraft 'Smart Bomb' Attack

Aircraft with Laser-Designator
Illuminates Target

Aircraft with LGB

Target-Seeking
Bomb's Trajectory

designator can be carried by helicopter or by front-line observers or infantry. 'Coded', in reference to the laser beam, means that impulses are generated, differing in length and/or spacing, according to a programme suitable for the particular type of operation. In the case of AGM-65, each missile can be tied by a unique coder to only one air or ground designator, thus allowing several to be launched independently in one vicinity.

This new type of bomb was used during the last years of the Vietnam war. The destruction of the Thanh Hoa bridge, a hundred kilometres from Hanoi, furnished proof of its great precision. This bridge was in a key position and repeated attacks had been made on it by US aircraft using conventional bombs without any success: it was destroyed on 12 May 1972 by a single laser-guided bomb. On 8 June of the same year, the Americans announced that fifteen strategically important bridges had been destroyed by laser bombs, thus considerably slowing down the passage of over 3000 North Vietnamese trucks taking supplies to the Vietcong.

Laser has also been used in missile guidance, endowing the missile with unprecedented accuracy. Another application of laser is in LADAR (Laser Detection and Ranging), a union of laser and radar

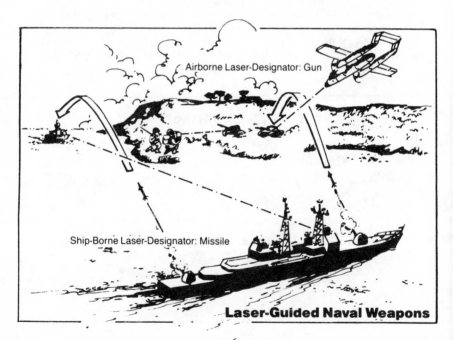

Airborne Laser-Designator: Gun

Ship-Borne Laser-Designator: Missile

Laser-Guided Naval Weapons

which is now used for many different purposes: to guide projectiles, including artillery shells, to position satellites, for accurate navigation—in short, in all operations where radar alone cannot give sufficient accuracy. Recently, the US Navy and Marines have carried out numerous experiments using laser to guide Naval artillery shells during amphibious operations. With this new system, every round fired hits its target, resulting in great savings on costly ammunition. It is an innovation which will certainly bring a new dimension to maritime warfare.

Low-Light-Level Television

It is well-known that acquiring information on the target to be attacked and, if possible, examining it and its surroundings is a basic requirement in military operations. Since time immemorial all sorts of means have been used to achieve such an important goal. Radar reveals the presence of an object but does not tell us what it is or what it is made of. We have already seen that infrared devices give us an idea of the nature of the target even in total darkness. Nowadays, with modern night vision techniques, it is possible to see almost as clearly in darkness as in daylight.

The most common technique used to improve human vision during periods of limited visibility is by image intensifiers used in Low-Light-Level Television (LLLTV). Image intensifiers work by amplifying ambient particles of light, such as the weak reflections of the moon and starts, which are always present in the atmosphere. The first image intensifiers were developed at the end of the 1950s but they were rather cumbersome devices and impractical for military use. However, interest in them was sustained due to their use by astronauts for making observations during space-flights.

Image intensifiers were first used for military purposes in 1965, since which time continuous improvements have been made to them. With modern versions, it is possible to see the light from a cigarette at a distance of 2 kms.

A further step forward was made in night vision techniques by combing television and image intensifiers, which led to the development of Low-Light-Level TV. This has the dual advantages of both intensifying the light level by a factor of at least six and also of separating the viewer from the image's source, thus making it unnecessary for the viewer to get used to the dark. In effect, with an LLLTV system, it is possible to intensify the weak light emitted by the stars so that one can see an area almost as clearly at night as in daylight. LLLTV is now widely used in aircraft and helicopters giving the pilot adequate night-vision for night-flying—including for take-off and landing as well as for navigation and in tactical operations at night or in poor visibility. Image intensifiers are also used in modern submarine periscopes.

Another system in widespread use is the airborne TV-aiming system also used at normal light levels whose design is very simple. These use special and very powerful zoom lenses which enable the operator to distinguish very clearly people walking down a street from an altitude of thousands of metres. The operator is able to view the target from the most appropriate altitude, depending on AA defences: as soon as the target has been framed in the screen, the operator launches the bomb or missile which, using the TV-camera to keep the target in view, is guided by radio-directed signals. Widespread use was made of TV-guided bombs during the last few years of the Vietnam war. In particular, aircraft operating from US aircraft carriers were equipped with the AGM-62 Walleye TV-guided glide-bomb which was particularly suitable for the destruction of targets which are difficult to hit, such as road and railway bridges.

Traditional artillery has also benefited from electro-optical inventions and, today, it is possible to correct the trajectory of a shell while it is in flight.

Electro-Optical Countermeasures

As had been the case with radar and infrared radiation, the widespread use of laser and LLLTV led to the development of appropriate countermeasures and counter-countermeasures. Since laser and LLLTV come within the field of electro-optics, such countermeasures are referred to as electro-optical countermeasures (EOCM). The subject is also called 'optoelectronics' but there has been a tendency recently to distinguish the two, limiting 'optoelectronics' to communications and information and 'electro-optics' to weapons systems and related countermeasures.

A laser beam is very highly directional and thus difficult to intercept. On the other hand, it can be easily deceived since it can operate only within very limited frequency bands. The most common deceptive technique is to use another laser which has similar characteristics but is much more powerful. This laser is beamed onto a point situated at a safe distance from the target to be protected. The 'laser searcher' fitted to the bomb or missile is thus deceived by the more powerful laser and directs the weapon towards this source rather than to the real target with the result that the bomb or missile impacts in a zone where it cannot do appreciable damage.

Passive CMs can also be used to counter lasers. These include reducing the effectiveness of the laser emission by using aerosol, smoke, chemical additives or other chemical substances which absorb or disperse its energy.

The problem of devising EOCMs to counter LLLTV and optical systems in general, including the human eye, is more complicated. One passive EOCM is 'optical-chaff' which works on the same principle as the tin-foil strips used against radar during World War Two and thereafter. Huge quantities of tiny sequins (paillettes) can be launched from an aircraft or ship under attack which, by reflection, can dazzle the TV camera of the enemy electro-optical search system.

It is worth mentioning countermeasures to the human eye which, in the conflicts in the Middle and Far East, has proved to be still one of the most effective aiming systems. One such system, exploiting the phenomenon of reflection, directs luminous energy towards the eye (through the same focussing lenses used for aiming) which interferes with the eye's vision, confusing or deceiving it as regards the position of the target. It is also possible to direct laser beams at the eye of the man aiming the weapon so that, by exploiting the lens through which the man is looking, the retina of the eye is damaged.

High Energy Laser Weapons

Although the laser has proved to be an effective guidance system for arms and munitions, efforts to develop a lethal laser weapon, a kind of 'death ray', have so far been unsuccessful; nevertheless, the Superpowers still doggedly pursue this aim.

In all probability, a portable anti-personnel laser 'death-ray' weapon could be developed without much difficulty and would no doubt be lethal. However, the fact remains that no one has yet come up with such a weapon: the reasons for this probably lie in the fact that it would be too easy to devise appropriate countermeasures which would neutralise the effectiveness of the weapon. Besides, it would be too expensive to use as an individual weapon: in theory, an ordinary mirror could be used to reflect the beam back to the sender, or the beam be evaded by hiding behind a wall or other structure, or by the use of aerosol systems; even better, clouds of dust or smoke could be created by throwing hand grenades, thereby blinding the optical aiming system and neutralising the effectiveness of the weapon.

In recent times, the two Superpowers have been directing their efforts towards the development of a 'high energy' laser, with a power of 5–10 megawatts, much more powerful than any laser now in existence. This weapon should, in practice, be able to produce and transmit high rates of energy through the atmosphere and concentrate them on high speed targets such as missiles and supersonic aircraft, boring through them or damaging their guidance systems as a result of thermal effects.

Air forces are particularly interested in the development of such weapons as a means of protecting bombers from air-to-air and surface-to-air missiles, especially in cases where traditional ECMs are unable to afford adequate protection during penetration of enemy air space. Naval forces, on the other hand, see a high energy laser weapon as a precious means of countering anti-ship missiles, including cruise missiles, and sea-skimmers which travel at very low altitudes. Finally, for ground forces, such a weapon could provide close AA defence against any type of attacker.

However, there are enormous problems to be overcome before such a weapon becomes a reality. The first problem involves transferring the high energy laser device from the gentle environment of the laboratory to the severe conditions of military vehicles or other platforms with the attendant constraints of power requirements, weight and space. Another obstacle to be overcome is atmospheric dispersion which is very strong at laser beam wavelengths. As with infrared radiation, the atmosphere greatly reduces the propagation and consequently the range of any laser, even high energy lasers. These problems could be partially solved by using the laser weapons at high altitudes or, better still, in outer space where there would be no absorption of energy.

In the United States several Boeing C-135 Stratolifter aircraft were converted for use as flying laser laboratories to conduct research on the use and installation of laser-weapons at high altitudes. These aircraft are equipped with high energy lasers and special aiming and tracking systems. One such aircraft disintegrated in flight on 6 May 1981, over Maryland while conducting secret experiments. Meanwhile, various testbed lasers have successfully shot down drones on several occasions, using several different types of laser-generator. Tests are carried out at the White Sands Missile Range where research is also done into the problems of the damage caused by laser to the metals (steel, aluminium, etc) of which targets are made.

The development of a high energy laser weapon by the US defence industry will take a long time and the possession of such weaponry by only one of the Superpowers would weigh heavily in the present balance of power. For these reasons, the Americans have already set aside increasingly large sums of money for the research and development of appropriate countermeasures to protect themselves, should a lethal laser-weapon eventually appear on the scene (see Chapter 23).

Very Low Frequency Weapons

After the death of Nikola Tesla in 1943, the United States, underestimating the technical and military value of his discoveries, authorised all his papers to be sent to Yugoslavia, which had requested their return. As soon as the papers arrived in Yugoslavia, they were secretly examined by Soviet intelligence specialists who immediately took possession of the most important studies and projects.

The Soviets were extremely interested in Tesla's research and, in recent years, have continued his research into the possibility of developing a new type of deadly weapon; a weapon which would no doubt have devastating effects but would be extremely difficult to develop at a practical level.

During his work on inductive coils, Tesla had also studied the possibility of transmitting electric energy from a distance without using normal conductors. He held that the Earth itself could be used as a conductor, as though it were a gigantic diapason able to emit vibrations on a particular wavelength. According to his theory, it was possible to make low frequency (6–8 Hz) transmissions through the Earth, using vertical-type waves emitted by the Earth itself.

In 1899, at Colorado Springs, USA, Tesla unveiled an inductive coil larger than any previously constructed and with it managed to light hundreds of lamps, at a distance of about 40 kms, by transmitting electrical energy through the Earth without using any electrical conductors.

He further developed his theory that a signal close to the frequency of basic resonance, which he estimated to be 8 HZ, could pass through the Earth and be picked up on the other side, the reason for this being that the propagation of the signal itself would be effected by vertical waves. Some US experts maintain that such a system could have been used by the Soviets to provoke seismic phenomena such as the earthquake which occurred in Peking at the beginning of 1977.

However, it must be pointed out that vast power and an enormous antenna would be necessary to provoke the Earth in such a way. In fact, to provoke seismic activity of the magnitude of the 1977 Peking earthquake, the Soviets would have had to use an antenna consisting of a 20 km copper plate which would surely not have escaped the notice of US reconnaissance!

The hypothesis that the Russians might have developed a low-frequency weapon on the basis of Tesla's theories is rather more feasible. Such a weapon would operate on a frequency of 8 Hz, which is very close to that of the human brain[2], and might therefore interfere with the workings of the mind just as ECMs interfere with radio and radar. It seems that impulse emissions on such a frequency can cause effects which range from drowsiness to aggressiveness. It has been reported that two special transmitters operating on this very frequency have already been built by the Soviets at Riga and Gomel. This resonance is also presently being studied in the United States for use in communications with submerged submarines.

This weapon, however, has a characteristic which Tesla did not foresee: it exploits a different type of resonance which is formed in the space between the surface of the Earth and the lower layer of the ionosphere. With such a system, Tesla's beam, besides being transmitted through the Earth, can also be transmitted around it.

The effects of electromagnetic fields on the human body have also been studied in the West. Thanks to the extremely sensitive measuring instruments now in existence, it has been discovered that the human brain and heart have magnetic activity. In the field of medicine, this discovery has given rise to magneto-encephalograms and magneto-cardiograms.

A great deal of interest has recently been shown in the biological effects of electromagnetic fields in the band of extremely low frequencies (ELF)—3 HZ–3 KZ. It is interesting to note that many electromagnetic atmospherical disturbances come within this frequency band and that these ELFs are similar to biological rhythms. A certain sensitivity to these frequencies has also been demonstrated by some animals. Reduced motor activity in birds has been noticed in the presence of electromagnetic fields of 1.75 and 5 Hz and an increase in such activity in fields of around 10 Hz. Also, many fish are sensitive to frequencies between 0.1 and 10 Hz.

Wewer and Altmann[3] report that electromagnetic fields in this band affect man's behaviour. The physiological basis for these observations involve the autonomous nervous system and endocrinic system. In short, an ELF weapon could conceivably be used to influence thought and thereby control the whole of mankind.

However, assuming for the sake of argument that such a weapon might one day be developed, it would not be difficult, given the low frequency and power, to devise effective ECCMs to protect our brains from this insidious danger which lurks behind the electromagnetic spectrum.

20

Micro-Conflicts, Limited Wars and Invasions

The Entebbe Raid

Besides major international crises, electronic warfare has also played a useful though little-known role in a number of smaller scale conflicts which have taken place in recent years, such as those caused by international terrorism.

A typical example of the use of electronic countermeasures in one such 'microconflict', as they are now called, was the Entebbe Raid when an Israeli commando force freed 102 hostages who were being held at Entebbe airport, situated about 20 kms from the Ugandan capital of Kampala. The series of events aroused worldwide interest and probably everybody remembers the story. However, few people are aware of the role played by electronic countermeasures in this enterprise or to what a great extent they contributed to the success of the Israeli operation.

On 27 June 1976, Air France flight 139, an A-300 Airbus, flying from Tel Aviv to Paris with 254 passengers on board, had just taken off from Athens when four terrorists belonging to the Front for the Liberation of Palestine hijacked the airliner, ordering the pilot to fly first to Bengazi and then to Entebbe.

The Israelis organised a special commando to free the hostages. They sent four C-130 Hercules transports and two Boeing 707s, escorted during the first part of their flight by F-4 Phantoms. No longer escorted, the Hercules flew in low over Lake Victoria and landed at Entebbe while the two Boeing 707s stayed in the air, functioning as operational command and control centres.

After a violent shoot-out, the commando freed the hostages and boarded them on one of the Hercules which immediately took off and flew to Nairobi where the wounded were disembarked, to be followed by the other three Hercules some thirty minutes later, after they had dealt with the remaining resistance, and sabotaged the Ugandan air force MiGs based at Entebbe.

The return trip to Israel meant an eight hour flight for the Israeli aircraft during which time they were exposed to Ugandan and Arab air force fighters. To avoid possible air attacks, the Israelis used the jammers on board one of the Boeing 707s to blind all airborne and local

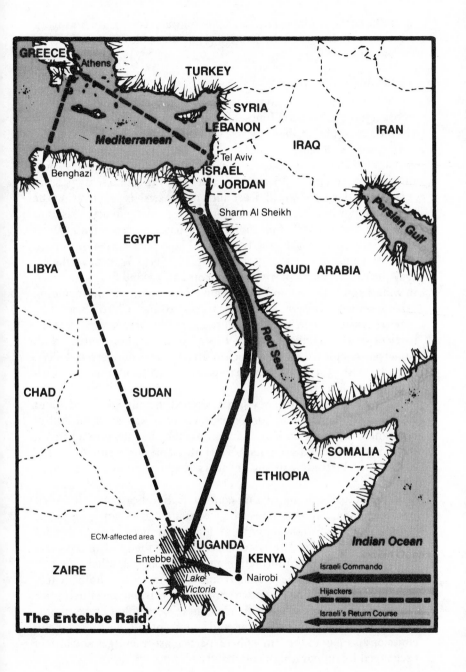

The Entebbe Raid

air traffic control radars. Thus, any possible intervention by Amin's air force was prevented and the Israeli aircraft were able to return to Israel undisturbed.

The Sino-Vietnamese War

After several weeks of continual border incidents, at 05.30 on 17 February 1979, twenty Chinese divisions, supported by hundreds of aircraft, tanks and artillery pieces, crossed the 1200 km-long Chinese border with Vietnam.

Although Chinese leaders repeatedly declared that they only intended to teach the Vietnamese a lesson, the Chinese aggression seriously endangered world peace and created serious problems for the two Superpowers. The Soviet Union had signed a military assistance pact with Vietnam only four months previously, so naturally, a Soviet armed intervention was greatly feared. The Soviet Union, uncertain whether to run the risk of starting a third World War, nevertheless took the precaution of placing all their air and ground forces stationed in Siberia on full alert and dispatching a naval formation, including missile-armed cruisers and destroyers, to the China Sea. The Americans also sent, as a precautionary measure, several aircraft carriers of the Pacific Seventh Fleet, to the troubled area. Both Superpowers placed all their nuclear attack forces on increased alert, starting with their submarines, carrying ballistic missiles with multiple nuclear warheads.

Meanwhile, what little news that reached the outside world about events along the Chinese border with Vietnam was, as usual, full of contradictions. The Chinese claimed that they had penetrated 80 km into Vietnam while the Vietnamese triumphantly proclaimed that the border was littered with the invaders' bodies and their destroyed tanks.

If the actual fighting of this conflict in Southeast Asia followed traditional, conventional lines, the electronic battles going on in the atmosphere were, on the other hand, highly advanced, with both sides using the most up-to-date systems provided by the two Superpowers to acquire all possible information and to spy on each other.

First of all, both the Russians and the Americans immediately launched supplementary photographic and electronic reconnaissance satellites for battlefield surveillance. These were able to take photographs of what was going on and intercept all electromagnetic emissions present in the atmosphere, particularly messages and orders exchanged by military high commands.

In order to have full reconnaissance coverage of such a complex zone as the Indochina peninsula and to keep an eye on the movements of enemy air and naval forces, the Russians also sent a number of Tu-95 maritime and electronic reconnaissance aircraft, equipped with the latest electronic surveillance equipment, to the Gulf of Tonkin. The Americans, on the other hand, sent a number of Grumman E-2C Hawkeyes to their strategic base at Okinawa, Japan. Specially designed and built for electronic reconnaissance at sea, these aircraft were to keep a close eye on the Russian naval units and to intercept all their electromagnetic emissions, both radio communications and radar signals. Expert analysis and interpretation of these emissions would provide the Americans with a clear picture of Soviet operational intentions. Thus, whereas no journalist, 'special correspondent', military attaché or secret agent was able to get hold of any reliable information about the Sino-Vietnamese war, the CIA and the GRU got all the information they needed via their eagle-eyed satellites and their electronic reconnaissance aircraft.

The Sino-Vietnamese war provided the Americans with an excellent opportunity to try out, under real conditions, their system of command, control and communications to which the nuclear defence and strike capabilities of the United States are entrusted. To his great surprise and consternation, President Carter learned that this whole system was now vulnerable to new Soviet space weaponry.

The Sino-Vietnamese conflict came to an end after a few months with no conclusive victory for either side. Both the Chinese and the Vietnamese officially declared that they had achieved their set objectives but it seems more likely that the conflict was brought to a halt by the high losses sustained by both sides.

A few months after the end of the conflict the People's Republic of China reported that many of their soldiers had been hospitalised in Canton for eye and brain lesions. The Chinese suspected that the Soviets had taken advantage of the conflict to try out a new secret weapon, most probably a high energy laser, using the Chinese soldiers as guinea-pigs.

Strategic Arms Limitation Treaties and the Iranian Crisis

The Sino-Vietnamese war broke out just when the USA and USSR were about to sign an agreement for the limitation of strategic arms (SALT-2). Great efforts had been made by both sides to overcome the two large stumbling-blocks of new US cruise and Pershing 2 missiles and Soviet Tu-26 *Backfire* bombers. Soviet opposition to the cruise missiles, which are equipped with a revolutionary guidance system called TERCOM (see page 229), comprising a computer and radio-altimeter and can thus fly with great navigational precision at very low level, arose from the fact that the Russians had not yet managed to devise any countermeasures, electronic or other, to them.

Opposition to US Pershing 2 missiles to be based in Western Europe also stemmed mainly from considerations regarding electronic warfare. Although these missiles have a shorter range than similar Soviet missiles installed in Eastern European countries, they were greatly feared by the Russians because of the difficulty of devising effective ECMs to their extremely sophisticated guidance system which, being of the inertial type linked to a special radar, is almost immune to jamming or deception (See page 230).

The Americans, on the other hand, expressed concern about the Soviet *Backfire* bomber, the official reason given for this concern being that there were more of them than the equivalent US F-111 which had only been produced in limited numbers. The EW version, EF-111, was considered indispensable for escorting bombers in deep penetration of enemy air space.

The signing of the SALT-2 agreements, which took place in Vienna at the end of June 1979, was a source of much controversy in the

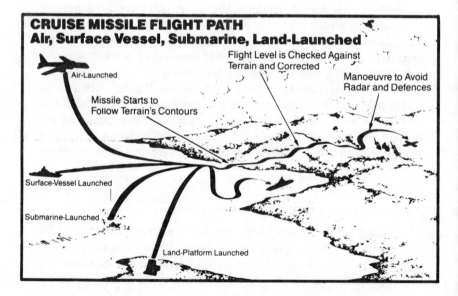

**CRUISE MISSILE FLIGHT PATH
Air, Surface Vessel, Submarine, Land-Launched**

Air-Launched

Missile Starts to
Follow Terrain's Contours

Flight Level is Checked Against
Terrain and Corrected

Manoeuvre to Avoid
Radar and Defences

Surface-Vessel Launched

Submarine-Launched

Land-Platform Launched

TERCOM Guidance

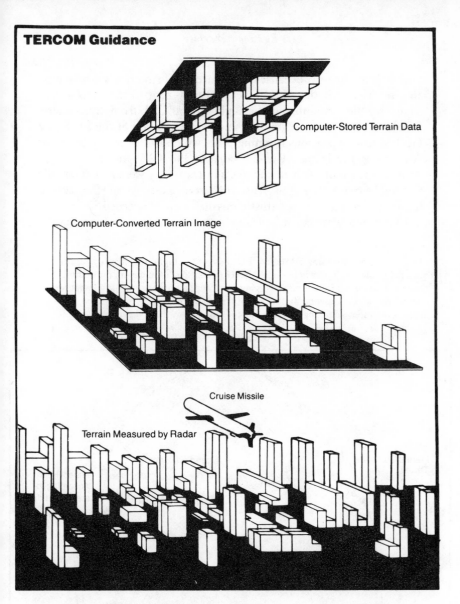

Computer-Stored Terrain Data

Computer-Converted Terrain Image

Cruise Missile

Terrain Measured by Radar

The TERCOM (Terrain Contour Matching) guidance system of the AGM-cruise missile is based on correlating the terrain-mapping data stored in the system itself with the terrain being overflown to locate the point fixes along the route to the target. In other words, the missile follows a pre-programmed flight path down to the target. The TERCOM's sole function is to compensate for course errors by comparing the terrain programme contour characteristics, stored as digital data in the computer 'library' of the guidance system, with the actual terrain contour characteristics as measured during flight by the radar altimeter. The maps are programmed using measurements collected by reconnaissance aircraft and satellites.

United States mainly because it was feared that America was no longer able to verify whether the Soviets actually kept to the agreements. Many people remembered the flight of the CIA from Iran where, following the Islamic revolution and the deposition of the Shah, the US had lost its precious listening posts which had been operating for years along the Iranian-Russian border. The CIA pointed out that, without the aid of the listening posts in Iran, they were no longer able to monitor whether the Russians were keeping to the agreed limitations as far as new ballistic missiles were concerned.

Every new Russian missile, before going into operation, has to

Flight path of a Pershing 2 MRBM. For most of the flight, the missile is under inertial guidance, but switches to RADAG (Radar Area Guidance) for homing onto the target. Only as it begins diving earthwards, does the missile's 'radar area correlator' kick-in, comparing the actual terrain below with a 'map' stored in its small computer, actuating external fins to bring the warhead down precisely on target.

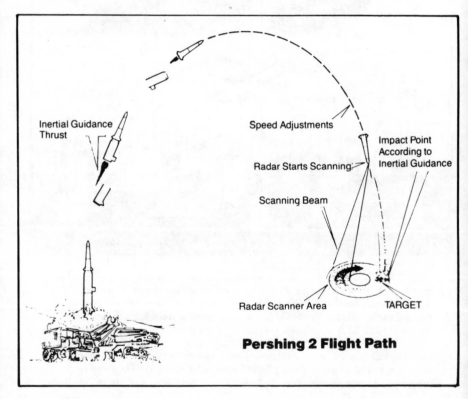

Pershing 2 Flight Path

undergo a series of test flights, on average more than twenty, which are carried out over a one-year period. During this period, it is absolutely essential to use radar guidance and radio command and control systems. This requirement enables the Americans to monitor the electronic characteristics of the missile and, consequently, to assess its operational performance. The CIA had always acquired such knowledge of Russian missiles via the ELINT listening posts in Iran. .

Many US Senators expressed grave concern about this 'intelligence vacuum and wanted to withold approval of the SALT-2 agreements until a project had been devised and developed to compensate for the loss of stations in Iran and to restore adequate electronic surveillance capabilities using other systems.

The Invasion of Afghanistan

The signing of the SALT-2 agreements was further held up by the Soviet invasion of Afghanistan. As had been the case in Iran before the Islamic revolution, there were evident shortcomings in US intelligence prior to the invasion of Afghanistan. A special inquiry was held to try to find out how the massive movement of so many Russian forces had escaped the attention of the various branches of the CIA.

CIA analysts had reported that only 15,000 Soviet troops were deployed within easy reach of the Afghanistan border. In fact, the number of troops massed in southern Russia was much higher and at least 85,000 actually took part in the invasion. Most of them landed at the airports of Kabul and Bagram, airlifted by 350 large transport aircraft, between the 24 and 27 December 1979, while four tank divisions and motorised infantry swept across the border. Moreover, many of these troops had been transferred some days previously from the Baltic to bases in central Asia and these movements had also escaped the notice of American intelligence-gathering satellites and other sensors because the Soviets had used deceptional tactics.

And yet, US listening stations had intercepted certain pre-recorded messages containing appeals to the Afghan people; these messages were for later re-transmission from Kabul at the moment of occupation. This demonstrated that the section of the US Intelligence Community which functioned best in the circumstances was radio interception (COMINT). This was further demonstrated by the fact that, while occupied Kabul was isolated from the rest of the world, only radio interceptions managed to provide any information about what was happening in central Asia. One must, therefore, deduce that

shortcomings were in the area of analysis and evaluation of information received, an activity which is usually carried out at the highest
national levels.

ECM and the Failure of the American Raid in Iran

When the US President, Jimmy Carter, made a surprise announce-
ment on the morning of 25 April 1980 to the effect that a secret
commando operation, undertaken during the night, to release the US
hostages being held in Teheran had failed tragically due to a technical
hitch, the whole world was gripped by alarm and apprehension at the
spectre of a nuclear war. When detailed explanation was later
furnished by the US Defense Department, the reaction was one of
consternation mixed with incredulity. How could it be, people asked
themselves, that the greatest military power in the world, master of the
most advanced technology, had had to call-off an operation of such
crucial importance for the American nation simply because a few
helicopters had broken down?

Military experts in several western countries were not convinced by
the official explanation and put forward the hypothesis that the real
reason for the US failure concerned ECMs actuated by the Soviets
during the operation. What, then, were the real reasons for the failure
of the operation and how far was electronic warfare responsible for this
failure?

The idea of a *blitz* operation, like that of Entebbe, to free the
hostages held in the US Embassy in Teheran had been under
consideration since November 1979, shortly after the Embassy was
occupied. However, it soon became apparent that an operation similar
to that carried out by the Israelis was out of the question since the
situation of the hostages in Teheran was completely different to that of
the hostages held at Entebbe.

Various plans, secretly drawn up by small groups of Pentagon
experts, were considered, the choice finally falling on a rather
complicated plan involving the use of helicopters. The choice of what
type of helicopter to use was not simple. Since they would certainly
take off from an aircraft carrier, it had to be a naval type, and the best
choice seemed to be the Sikorsky S-65, variants of which—the
CH-53A Sea Stallion and the RH-53D—were operated by the US
Navy and the Iranian Navy, and thus, when the US helicopter
approached the Embassy to release the hostages, the Iranians might
think that it was one of their own machines.

The operation was divided into two stages. In the first stage six C-130 Hercules, carrying the ninety men of the commando and a large amount of fuel, would take off from an Egyptian airport, fly over the Red Sea, skirt around the Saudi Arabian peninsula and land at an old, unused salt airstrip in the desert of Dash-el Kevir near the Iranian town of Tabas, about 450 km from Teheran. At this landing field, designated 'Desert One', a rendezvous was fixed with eight RH-53 helicopters from the aircraft carrier *Nimitz*, cruising in the Gulf of Oman. The purpose of this rendezvous was to refuel the helicopters after their 500-mile mission and to transfer the commando to the helicopters.

In the second stage of the operation, which never took place, the commando on board the helicopters had to transfer first to a secret location in the mountains and thence to Teheran where, with the help of infiltrated secret agents and possibly sleeping gas, they would penetrate the Embassy, rescue the hostages and transport them to safety. Provision was made for instantaneous satellite relay communications between the commando and the Pentagon.

As in all operations of this nature, surprise and speed were absolutely essential pre-conditions for success. It was evident from the early planning stage that, with regard to these two elements, there were two particularly crucial problems to be solved: first, to avoid being discovered by enemy radars or other EW equipment and, secondly, to avoid any armed conflict with the Iranians.

To achieve the first aim, detailed EW plans were devised to be put into operation both before and during the mission itself. First, the Americans began to intercept all radio communications between the Iranian Embassy in Washington and the Foreign Ministry in Teheran to provide the Pentagon with information that might be of use in the planning of the mission. Secondly, so as not to arouse the suspicion of the numerous Soviet warships cruising in the Gulf of Oman and the Arabian Sea, nightly exercises were carried out by US air and naval forces using helicopters which frequently flew right up to the Iranian coast. Every night, other US ships, positioned away from the aircraft carrier, *Nimitz*, launched from special rockets, false targets consisting of radar reflectors to simulate the presence of helicopters in flight and thus occupy and confuse the Soviet radar operators.

US ships and aircraft in the zone transmitted false exchanges of radio messages each night so that, on the night of the actual operation, there would be no increase in radio traffic to arouse the suspicions of the ever-present Soviet spy-ships which systematically intercepted all radio-telegraphic traffic. In short, the aim was to make the Russians think that the departure of the eight RH-53 helicopters on the night of the actual mission was just another routine nocturnal exercise.

The operation was code-named 'Eagle Claw' and this name was used in all radio communications regarding the operation. It was essential to ensure maximum security of communication between the Pentagon, the aircraft carrier *Nimitz*, aboard which a special command structure was established to be in charge of the operation, and the forces taking part in the operation. To achieve this, the United States, at the beginning of January, secretly launched two communications satellites which used new transmission and coding techniques which rendered their communications almost completely immune to jamming and deciphering. At the same time, a reconnaissance satellite was launched into geostationary orbit over the Indian Ocean to ensure full photographic and electronic reconnaissance coverage of the area. In order to provide warning of any aircraft approaching the US C-130s and helicopters flying towards Iran, several USAF Boeing E-3A Sentry AWACS would be present: these were each equipped with a very long-range radar capable of detecting aircraft or helicopters at a distance of several hundreds of miles.

The most difficult problem to solve was that of penetrating and operating in Iranian air space undetected. Luckily, the Iranian air defence radar system had been designed and built by US industry some years previously and so, aided by the new electronic reconnaissance satellites, a 'blind' radar corridor was identified between the area covered by two Iranian radars through which the US aircraft and helicopters stood a good chance of flying undetected.

Both the C-130s and the RH-53s were equipped with electronic jammers to be used along their route to jam or confuse any communications between Iranian fighters and their ground control centres. Two C-130 Hercules were also armed with rapid-fire 7.62 mm machine-guns to support the assault on the US Embassy if necessary.

Finally, about 200 attack aircraft from the aircraft carriers *Nimitz* and *Coral Sea*, would be ready to intervene if the commando got into difficulties.

Two weeks prior to the date fixed for the raid, a C-130 was sent out at night through the 'blind' corridor to land at 'Desert One'. Its mission was to check the feasibility of penetrating Iranian air space undetected and to take earth samples of the salt desert for analysis to ensure that the heavy aircraft and helicopters could safely land there.

The actual operation began on the evening of 24 April when the six

C-130 Hercules took off from the military airport of Khena in Egypt. Later, at 19.30, the eight RH-53 helicopters took off from *Nimitz*, cruising in the Gulf of Oman.

To confuse the radars of the Soviet ships in the area, numerous false radar targets were launched from other US ships, not only in the Gulf of Oman but also in the East Mediterranean. Further confusion was created on the Soviet radar screens by the presence of numerous Israeli warships which (perhaps by sheer coincidence) had decided to carry out air and naval exercises that very night!

To avoid detection, the C-130s flew very low, first over the Red Sea and then the Gulf of Aden. Here, they had to switch on their jammers to blind Soviet radars installed in South Yemen and on the coast of Eritrea. After a brief stop at Masirah airport in Oman for refuelling, they then flew on to 'Desert One'. The helicopters took off from *Nimitz* and headed directly for the Iranian coast. They also flew at a low altitude along a route far from populated areas to avoid being detected.

To aid them in their ground-hugging flight, the C-130s and RH-53s were all equipped with the most advanced and accurate navigation systems then in existence, including INS (Inertial Navigation System) and Omega (a highly accurate very low frequency radionavigation system), as well as night vision devices.

The helicopter formation had hardly covered a third of the distance between *Nimitz* and 'Desert One' when a warning light on the instrument panel of Helicopter No. 6 lit up, indicating a risk of main rotor failure, a rare but potentially very serious occurrence.

The helicopter immediately landed near a small lake over which it had been flying. In accordance with pre-arranged procedures for maintaining radio silence, the last helicopter of the formation, No. 8, automatically followed No. 6 to provide assistance. On landing, a quick examination of the rotor blades confirmed the gravity of the situation. The commander decided to abandon the helicopter and he and his crew boarded Helicopter No. 8 which immediately took off and headed for 'Desert One'.

Then, the formation suffered another setback when they ran into a sudden, violent sandstorm. Visibility dropped sharply to almost zero and it became impossible for the helicopter crews to see the other helicopters in the formation even with their sophisticated night-vision devices.

At 21.30, the first C-130, carrying the men who were to set up the refuelling base, landed at 'Desert One', but, after only a few minutes, there was another unexpected setback. A bus carrying about forty Iranian civilians suddenly appeared coming along a dirt road which passed near the landing strip. The US officer in charge of the first group of men immediately stopped the bus but, not knowing what to do with the passengers, radioed the *Nimitz* for instructions. He was told to hold them but keep them well away from the refuelling zone. To crown it all, a few minutes later a tanker and a truck appeared, their headlights on, heading towards the airstrip where the other C-130s were just landing. The drivers of the tanker and the truck came to a halt when faced by this unusual spectacle and then fled into the darkness.

Meanwhile, the helicopter formation was still fighting its way through the sandstorm. Just before midnight, the crew of Helicopter No. 5 experienced gyro failure rendering unreliable both navigation instruments and more importantly, stability reference systems, the loss of which made it extremely difficult to follow the route accurately and keep the helicopter straight and level. The formation was just then approaching a chain of mountains reaching up to 10,000 feet lying across the route to 'Desert One'. The commander of Helicopter No. 5 therefore had to make a very difficult decision: whether to fly along the valleys as planned or over the mountain chain. The first option seemed extremely dangerous given the faulty gyros, while flying over the mountains would expose the helicopter to the search radars of both Iranian and Soviet defence systems. Probably for the latter reason, the pilot decided to fly back to the *Nimitz*.

At 00.30, the Pentagon received the news via satellite that No. 5 was returning to the aircraft carrier. The fact that now two helicopters, because of breakdowns, were no longer able to take part in the operation caused great consternation in Washington. But it was too late to replace them; in fact, no contingency plans had been made for replacements!

Shortly afterwards, a warning light came on in the cockpit of Helicopter No. 2, this time indicating a drop in pressure of the secondary hydraulic system which regulates the pitch of the rotor blades and, consequently, the speed of the helicopter.

Between 00.50 and 01.40, the six remaining helicopters finally landed at 'Desert One'. Examination of No. 2's hydraulic system showed that the fault was too serious for the helicopter to be used in the mission.

At this point, according to the official version of what happened, the three commanding officers (of the helicopters, the commando and the

'Desert One' base) conferred and came to the conclusion that the mission could not be carried out with only five helicopters. They they radioed command on board *Nimitz* who in turn transmitted a report to Washington suggesting that the operation be cancelled. President Carter was in agreement and gave orders for the aircraft and helicopters to return to their bases.

Back at 'Desert One', during the confusion of the preparations for their return, an RH-53 helicopter and a C-130 collided, causing a fire in which eight US soldiers died.

This is the official version of events but is not very convincing, not only for the reasons given but also because of more specific considerations.

First of all, it is difficult to understand how the Americans, who had had hundreds of helicopter experts based in Iran for several years during the Shah's regime, could have so underestimated possible technical problems arising from a 500-mile flight of delicate machines in a desert area where sandstorms are hardly a novelty. It is also difficult to understand why, for such a complex operation, it had not been deemed necessary to have reserve helicopters on board either *Nimitz* or another ship ready to replace any helicopter which might become unserviceable.

It must also be pointed out that, since each RH-53 helicopter could carry up to fifty-five people, the five remaining helicopters would probably have been adequate to rescue the fifty-one hostages and various secret agents. It has been reported that many of the commando's officers wanted to go ahead with the mission using the five remaining helicopters and tried to convince those who held that five helicopters were insufficient. Since there was no single officer in overall command, heated discussions ensued which ended only when Carter's orders to return arrived at 02.30. After the tragic collision, which happened at 03.18, there was great confusion at the base and this could also explain the almost incredible fact that operational plans and various top secret electronic devices were left on board the abandoned helicopters.

As we have already seen, Iranian territory, particularly the area near the Russian border, had been in the Shah's time a 'hot' zone from the point of view of EW: on the one hand, there were the Americans trying to intercept the emissions of Soviet radars during test-launches of new missiles at Tyuratam in Kazakistan; on the other hand, there were the Russians trying to prevent such interceptions by ECMs and ECCMs. With the advent of the Islamic revolution, the US surveillance establishments had been dismantled whereas Russian systems had remained intact and perhaps had even been strengthened because of the crisis in the Persian Gulf. It must also be borne in mind that the route followed by the US helicopters is well within the range of Soviet air defence radars installed along the Soviet/Iranian border.

Thus, from the point of view of EW, the first hypothesis one can make is that the Soviets, having located the helicopter formation by radar or other electronic means, jammed the American's radio communications and navigation systems, thus hindering navigation and preventing the exchange of orders and reports between the helicopters themselves, the commando force and secret agents whose help was needed for them to reach the US Embassy in Teheran.

Another hypothesis which has been put forward is that Soviet reconnaissance satellites intercepted the radio and radar transmissions of the US aircraft and helicopters and followed their movements over Iranian territory. Since, to reach 'Desert One', the US formations had to fly in the direction of the Afghanistan border, the Russians may have feared an attack on their forces in Afghanistan. Brezhnev might have got Carter on the famous 'hot-line' to dissuade him from undertaking any military operations in that part of Asia.

A third and perhaps more credible hypothesis is that, convinced that the numerous radio transmissions made by the helicopters during their various difficulties and by the commando following the unexpected appearance of the bus-load of Iranians had been intercepted by unfriendly stations, the Pentagon feared that the operation had lost the element of surprise which was indispensable for its success. Consequently, Carter, fearing a direct encounter with the Iranians, decided that it would be wiser to order the force to withdraw before it was too late.

However, according to various statements made by the Americans, neither Navy surveillance systems nor the E-3A AWACS flying over the zone had intercepted messages or signals indicating that Soviet radars, including those in Afghanistan, had discovered the presence of hostile or unidentified aircraft flying low over Iranian territory. Moreover, neither the Russians nor the Iranians have to this day made any declaration claiming responsibility for the failure of the American raid.

Regarding control of radio-electric emissions, given the secrecy required for such a mission, the exchange of messages with Washington and the aircraft carrier should have been avoided. Even though the ultrasophisticated transmission and coding techniques employed are highly resistant to interception (spread spectrum, pseudo-random-noise and so on), in an area so crowded with SIGINT platforms—ground stations, satellites, ships and aircraft—as in the Middle East, there was always some risk that messages could be intercepted and deciphered.

International Crisis Management

In addition to those already described, crises of varying gravity have taken place and, in fact, continually take place all over the world—recently, for example, in Central America, the Horn of Africa, Cambodia, Angola, Namibia and the Persian Gulf as a result of the war between Iraq and Iran, to name a few. Crises frequently occur in areas where, for political, military or geographical reasons, the acquisition of information on the local situation via normal channels is difficult if not downright impossible.

However, the two Superpowers have a vested interest in every international crisis. Directly or indirectly, each crisis affects the strategic and military balance of power, for example, due to concern for oil supplies, between the two alliances, NATO and the Warsaw Pact, with the omnipresent danger of a direct confrontation with the consequent risks of nuclear catastrophe. The aim of both Superpowers is to reap the maximum advantages from the crisis without committing their own forces to actual warfare and, at the same time, to prevent the other Superpower from gaining any advantage from the resolution of the crisis.

They must also ensure that they do not lose control of critical situations and thus risk becoming involved by error or by chance in a nuclear conflict. Where international agreements for the maintenance of a certain balance of political and military power exist, each side is constantly on guard in case the other side should try to cheat. Therefore, it is absolutely vital for world powers to rapidly acquire and accurately evaluate all possible information regarding every international crisis in order to actuate appropriate countermeasures (political, military, electronic and so on).

There were serious shortcomings in this activity, as carried out by US Intelligence services on the eve of the Iranian crisis and the invasion of Afghanistan which aggravated the dramatic problems facing the United States. These served as a reminder of the need for instruments capable of following the activity of potentially hostile forces in areas involved in international crises, and of detecting buildups of armour and troops along the borders of threatened countries and of following their movements day and night. Apart from satellites, this type of surveillance can also be carried out very effectively by aircraft and ships equipped for SIGINT.

This new task, which is both strategic and tactical and involves both air and naval forces, has been appropriately named 'Crisis Management'. Such surveillance missions must be carried out from a safe distance, never involving flying over the 'hot' zone but rather flying around the edges of the zone by night and day, employing electronic, photographic and IR equipment which can operate from great distances. Among Western aircraft equipped for crisis surveillance are the latest US reconnaissance aircraft, the TR-1 and EF-111A, the Boeing E-3A Airborne Warning and Control System (AWACS), and the EA-6B Prowler, E-2C Hawkeye, S-3A Viking and OV-1 Mowhawk, and the British BAe Nimrod and others. Among Soviet aircraft equipped for the same function, besides the ever present Tu-95 *Bear*, the Tu-16 *Badger-H* and the MiG-25R *Foxbat* which have already been mentioned, there are also the ECM-escort Yakovlev Yak-28 *Brewer-E*, the Tu-22 *Blinder* and the Tu-26 *Backfire-B*, not to mention the Tu-126 *Moss* which has a long-range radar system very similar to that of the US AWACS.

Many of these aircraft have been used during the course of the conflict between Iraq and Iran, in conjunction with the huge naval forces deployed in the Gulf of Oman, to follow events in that 'hot' zone. In particular, the Americans use four AWACS, based in Saudi Arabia, to monitor the entire air space of the Middle East and so avoid surprise attacks on their own naval forces operating in those seas.

The Falklands Conflict

On the night of 1/2 April 1982, a few miles from Port Stanley, the capital of the Falkland Islands, ninety marines transferred from the Argentinian destroyer *Santissima Trinidad* to landing craft and headed towards the coast. On landing, the commando split up into two groups: the first, composed of thirty men, headed for the British Governor's residence in Port Stanley, while the other, composed of sixty men, headed for the Royal Marines barracks. This was the first stage of 'Operation Tom'—the military occupation of the Falkland Islands—or the Islas Malvinas—by Argentina.

The thirty men of the first group, led by a lieutenant commander, met with strong resistance from the Royal Marines at the Governor's residence. Their officer himself was killed but the Argentinians had overwhelming superiority in firepower, and the Governor felt that he had no option but to order the Royal Marines to surrender.

Meanwhile, the main body of the Argentinian invasion force had landed on the Islands, arriving on board the corvettes *Granville* and *Drummond*, a submarine, various other ships, and several C-130 Hercules and Fokker transport aircraft. They soon overcame resistance and raised the Argentinian flag over the disputed islands.

Three days later, a British Task Force set forth from Portsmouth to regain the Falkland Islands—'Operation Corporate'—while desperate attempts to solve the problem diplomatically were being made. Nothing happened for almost a month while the British Task Force continued to sail towards the Falklands, their pace seeming to indicate more a diplomatic gesture than serious military intentions. The whole world followed the anachronistic affair with great curiosity and incredulity.

Meanwhile, the Soviet Union had begun to send a series of spy-satellites into orbit to keep an eye on events in the South Atlantic. They had also hastily sent a number of Tu-95 *Bears* and the usual spy-ship disguised as a fishing-trawler to maintain surveillance of the British Task Force. The largest Soviet naval aircraft the Tu-95 *Bear* has been produced in several versions. the *Bear-D* is used for maritime reconnaissance and surveillance missions. For surveillance in the South Atlantic, they usually operate from a Cuban-controlled military airbase in Angola. Such aircraft also have ELINT capabilities.

The first Soviet satellites were sent into orbit on 31 March, two days before the Argentinian landing. They were the satellites Cosmos 1345 and Cosmos 1346 and their main tasks were, respectively, to intercept radar emissions (ELINT) and listen to and record radio communications (COMINT). On 2 April, a photographic reconnaissance satellite, Cosmos 1347, was also sent into orbit: this would drop capsules containing exposed film each time it passed over a fixed point in the USSR. Between 16 and 23 April, the satellites Cosmos 1350, 1351, 1352 and 1353 were launched to replace those in orbit and to continue their surveillance activities. On 29 April, the satellite Cosmos 1355, specialising in oceanic surveillance, was launched.

The Russians then sent into orbit other satellites (Cosmos 1356, 1357, 1364, 1366, 1367, 1369 and so on) for the specific purpose of monitoring operations in the Falklands. Some of these satellites were able to establish the positions of all ships present in the South Atlantic and to take photographs which were simultaneously transmitted to Russian ground stations for interpretation.

The Americans had already been following, via satellite, the course of events in Argentinian ports and had, in fact, warned the British of the imminence of a landing in the Falkland Islands. Their surveillance was not limited to the South Atlantic, however. According to unofficial sources, the Americans used their mammoth National Security Agency (NSA). The Agency has satellites for communications (COMSAT), very well-equipped ground intercept stations and decoding centres employing huge computers specially designed by IBM. NSA used these assets to intercept Argentinian communications and break the codes, thus enabling them to furnish the British with valuable information regarding the deployment of Argentinian forces in the Falkland Islands and the movements of Argentinian ships.

It is not known whether the Russians managed to break the Royal Navy's tactical codes in time to be of use to them. However, there is no doubt that the British ships kept their radio transmissions to the absolute minimum when a Russian satellite was passing overhead.

On Sunday, 2 May dramatic news arrived from the far reaches of the South Atlantic that the British nuclear-powered submarine *Conqueror* had torpedoed the Argentinian cruiser *General Belgrano* off the coast of Patagonia. The *General Belgrano* was an ex-US Navy World War Two-vintage 13,645 ton 'Brooklyn' class cruiser, USS *Phoenix*. The cruiser had been sailing towards the Task Force but was still outside the 200-mile Total Exclusion Zone which the British had declared must not be entered. She carried no anti-submarine warfare

(ASW) equipment but was escorted by two smaller ASW ships which nevertheless lacked modern anti-submarine equipment. Like the Argentinian ships which had taken part in the landing operation, the *General Belgrano* had used radar and radio rather imprudently, perhaps not understanding that all her electromagnetic emissions were being intercepted by the Americans who allowed access to all such information to their NATO ally, Great Britain.

It was not difficult for the British submarine, its nuclear propulsion enabling it to travel fast underwater, to reach the old Argentinian ship and place itself in a suitable position for firing. However, before launching torpedoes, the commander of the submarine, Commander Christopher Wrexford-Brown RN, rightly decided to radio London for instructions. The British Prime Minister, Mrs Margaret Thatcher, advised that the Argentinian ships constituted a clear threat to the approaching Task Force, gave the order to the submarine's Captain to torpedo the enemy ship.

At 16.00, the first torpedo hit the cruiser under the water-line near the aft engine-room. The electricity supply was immediately cut off and the ship was plunged into total darkness. Three seconds later, another torpedo hit near the bow. At 16.07, the ship, fatally hit, began to list so badly that, 15 minutes later, the Captain, Bonzo, gave the order to abandon ship. The sea was very rough and the two escort ships, fearing that they would also be hit, rapidly withdrew. Rescue operations were difficult and about 400 Argentinian sailors lost their lives in that controversial, widely criticised, engagement, whose operational necessity was called into question. At 17.00 on 2 May the cruiser went down with her flag still flying.

From an operational point of view, there was nothing surprising about the sinking of the *General Belgrano*. The southern force of an Argentinian naval pincer movement designed to encircle the British Task Force and bring it to battle, she constituted a serious threat to the Royal Navy and the embarked land forces, whether or not she entered the Total Exclusion Zone, the declaration of which apparently limited British offensive operations. Interestingly, the commander of Argentine Naval forces in the South Atlantic asserted (on BBC television in 1984) that the sinking of the *General Belgrano* outside the Total Exclusion Zone was a legitimate and necessary act of war and one he would have taken.

What is surprising about the episode is the ease with which *Conqueror* despatched the cruiser. As she lacked anti-submarine equipment and an adequate escort, it was virtually inevitable that she would be sunk. However, *Conqueror* had been tracking the cruiser for many hours, and, moreover, sank her with World War Two-vintage (Mark 8) acoustic torpedoes, because her commander distrusted the modern, wire-guided Mk 24 Tigerfish type as, although they employ highly sophisticated equipment, the total system reliability is inadequate. If *General Belgrano* could be sunk with such comparative ease by torpedoes whose basic design was some forty years old, fired from relatively close range, the question is raised of what warships can do against deadly modern homing torpedoes fired from long range by fast nuclear-powered submarines.

Countermeasures can be used, with a certain measure of success, even against modern torpedoes: these must, of course, be underwater acoustic measures since electromagnetic waves cannot be propagated in water as effectively as can sound waves. The classic method of avoiding acoustic torpedoes, devised during World War Two, was by means of acoustic deception; a noise-generator, transmitting the same kind of noise as that produced by the ship, but louder, was towed by the ship one sought to protect, and the torpedo was thus diverted towards this false target rather than the ship.

Naturally, devices used today are quite different from those used in World War Two. Technological progress in this field has led to the development of new systems for deceiving or destroying torpedoes which are computer-controlled and completely automatic, for example, anti-torpedo torpedoes. Since modern torpedoes, including wire-guided torpedoes, are equipped with an acoustic homing system for the final phase of the attack, acoustic measures, countermeasures and counter-countermeasures have been devised. These are constantly being refined, each side trying to out-do the other, just as in the electromagnetic conflict in the airwaves.

The Argentinian ships were no match for the British nuclear submarines and the sinking of the *General Belgrano* gave clear proof of which country ruled the waves!

The problem of countering nuclear submarines faces the largest navies, including those of America and Russia in rather greater degrees. Great efforts have been dedicated to the construction of chains of computer-controlled underwater acoustic sensors, which are able to detect submarines long before they are in a position to attack.

The Argentinians soon retaliated for the sinking of the *General Belgrano*. On 4 May 1982, an Argentinian PV2 Neptune maritime patrol aircraft sighted a British naval formation, consisting of one large ship and a smaller one, at a distance of about 70 miles to the south-east of the Falklands. These were the aircraft carrier HMS *Hermes* and the Type 42 destroyer HMS *Sheffield*; the latter was operating as a radar picket about 20 miles from the larger ship. Two Super Etendard strike aircraft, both armed with AM-39 Exocet missiles, were immediately sent out by the Argentinian high command to attack the two ships.

The two aircraft flew low, skimming the tops of the waves, to avoid radar detection. The P2V Neptune which had sighted the British naval formation guided the two Super Etendards to their targets and also controlled their brief climb to enable target acquisition by their nav-attack radar. At a distance of about 25 miles from the point where the British ships had been sighted, the two Super Etendards climbed rapidly to 500 feet, briefly switched on their radars to locate the two naval targets and thus programmed the Exocet missiles' computers, and then returned to their previous low altitude. Weather conditions were bad with fog which reduced visibility to a quarter of a mile. At a range of about 23 miles each aircraft launched its missile and then headed back to base, having 'seen' the targets only by radar.

However, during those few moments in which the two Super Etendards had had their radar switched on, a British ship in the area had intercepted their emissions. The interception was immediately relayed to all ships of the Task Force, including *Hermes* and *Sheffield*. The *Hermes*, the formation's Air Defence Control Ship, identified the intercepted emissions as probably coming from Argentinian Mirage III interceptor or tactical strike aircraft and not from Super Etendards. This error of judgement probably meant that time was wasted in discussion and that the danger was seriously underestimated. Besides, the fact that the two aircraft had already turned back seemed to indicate that they had decided not to attack. Moreover, the British considered that the Argentinians were not as yet trained to launch Exocet missiles from Super Etendards. For all these reasons, the British did not attach due importance to the intercepted radar emissions.

At that precise moment, *Sheffield* was transmitting and receiving messages via satellite,[1] an operation that requires all equipment emitting electromagnetic energy to be switched off to avoid interference with the satellite communications systems: this was probably a major reason why the ship's radar did not detect the enemy aircraft in time. Moreover, *Sheffield*'s ESM (Electronic Support Measures Systems) did not pick up the radar emissions of the missile, either, which is strange as the missile's radar seeker activated at a distance of about 10 kms from the target.[2]

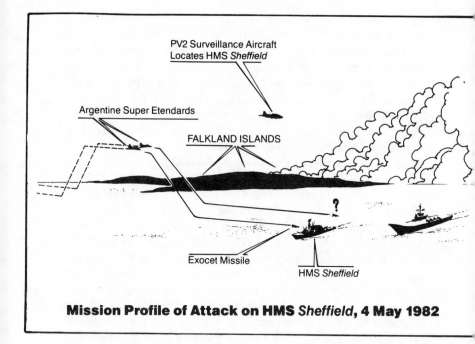

PV2 Surveillance Aircraft
Locates HMS *Sheffield*

Argentine Super Etendards

FALKLAND ISLANDS

Exocet Missile

HMS *Sheffield*

Mission Profile of Attack on HMS *Sheffield*, 4 May 1982

On the other hand, there was a dense electromagnetic environment in the zone, coming from radio communications IFF and radar equipment on board the British warships and the numerous merchant ships which had been sent to the Falklands to provide logistic support for the Task Force.

Meanwhile, the two undetected sea-skimming missiles, travelling at close to the speed of sound, covered the distance separating them from their targets in around two minutes. Just four seconds before impact, a look-out on the *Sheffield*'s bridge saw one of the missiles just in time for the Captain to order the crew to take cover. The missile hit the *Sheffield* amidships, about 6 feet above the water-line and penetrated the Operations Room. The other missile ended up in the sea, probably due to malfunctioning of its guidance system or, perhaps, for some other reason. The missile which hit *Sheffield* penetrated the hull, causing a terrible fire in which twenty men died and twenty-four were injured. Fed by the residual fuel of the missile, the fire raged for some time, like a huge torch. Electric cables, the ship's nervous system, also caught fire and the ship's pressurised ventilation system allowed the fire to spread throughout the ship. The hull was white hot where it had been hit and the crew could hardly move because of the thick smoke which had filled the whole ship making it difficult to breathe. Nevertheless, they desperately fought the fire for four hours, trying to save the ship, but when the flames encroached dangerously near the missile and combustible stores the Captain gave orders to abandon ship.

But *Sheffield* did not explode or sink immediately. She was put in tow with the hope of getting her back to Britain; however, after six days at sea, the badly damaged and burnt ship finally went down on 10 May, during a severe gale. It has been suggested that, although the Exocet may not have detonated, *Sheffield*'s bottom plates were severely damaged by blast and fire.

The *Sheffield* was the first of a class of twelve destroyers designated Type 42: The ship design had been criticised for lack of both defensive and offensive armament; in fact, these 4100–4700-ton ships, total armament is comprised of one twin-Sea Dart—SAM-launcher, one 4.5 inch gun, two 20 mm Oerlikon cannon and six ASW torpedo tubes, plus one Lynx ASW helicopter. Modern warships protect themselves

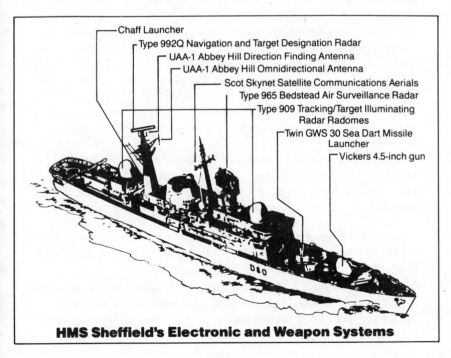

Chaff Launcher
Type 992Q Navigation and Target Designation Radar
UAA-1 Abbey Hill Direction Finding Antenna
UAA-1 Abbey Hill Omnidirectional Antenna
Scot Skynet Satellite Communications Aerials
Type 965 Bedstead Air Surveillance Radar
Type 909 Tracking/Target Illuminating Radar Radomes
Twin GWS 30 Sea Dart Missile Launcher
Vickers 4.5-inch gun

HMS Sheffield's Electronic and Weapon Systems

against anti-ship missiles by using either 'soft-kill' weapons (Electronic Counter Measures) or 'hard-kill' weapons such as anti-missile missiles, for example, the British Sea Wolf, and very rapid-firing guns.[3]

In terms of 'hard-kill' weapons, the *Sheffield*'s Sea Dart SAM system had an anti-missile capability, but had a shorter range than the Exocet. Moreover, the British had no AEW aircraft, having abandoned them with their aircraft carriers, so any warning of attack was limited to 'line-of-sight' detection by shipborne radars—in fact, *Sheffield* was acting in the EW capacity. This meant that the Super Etendards could launch their missiles from outside the range of Sea Dart missiles, this favourable situation being referred to as 'stand off'[4] capability. The aircraft were in no danger of being shot down, and their missiles dropped to sea level for their attack. *Sheffield* was not equipped to counter a sea-skimming threat, her equipment having been developed before missiles with this capability entered service, and, moreover, Britain had not expected to be at war against NATO weapons. The only possible 'hard defence' that *Sheffield* could attempt was bursts from her 20 mm cannon, which would have been totally ineffective against the tiny target of the oncoming Exocet missile.

Analysis of 'soft-kill' weapons on board *Sheffield*, in the absence of any firm official documentation on this sensitive subject, necessitates examination of the ship's superstructure, masts and antennae, all of which can be clearly seen on photographs. A photograph of the ship taken after she had been hit by the Argentinian Exocet (see plates) shows, on the mainmast, the antennae of an ESM equipment called UAA-1 Abbey Hill, a well-known ESM receiver produced by the British firm MEL Equipment Co. Ltd and introduced into the Royal Navy in 1973.

UAA-1 is a radar intercept and direction-finding system designed for use in surface ships operating in a dense radar environment. Developed in 1965–70 and, therefore, limited to the technology of that period, the two main operational functions of the system are:

1. early warning of the presence of radar-type transmissions from ranges beyond the horizon constituting a top-priority threat to the ship

2. general surveillance of the electromagnetic spectrum, intercepting, analysing and identifying radar-type transmissions within the frequency band 1-18 GHz together with an indication of the direction of arrival.

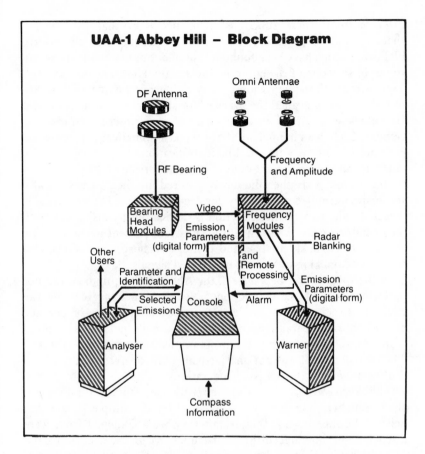

UAA-1 Abbey Hill – Block Diagram

Omni Antennae

DF Antenna

RF Bearing

Frequency and Amplitude

Bearing Head Modules

Video

Frequency Modules

Emission Parameters (digital form)

Radar Blanking

Other Users

and Remote Processing

Parameter and Identification

Console

Emission Parameters (digital form)

Selected Emissions

Alarm

Analyser

Warner

Compass Information

Block diagram of the main units of the UAA-1 Abbey Hill system. Signals are fed to the bearing receivers, located in a unit immediately below the aerial, and thence to the Remote Processing Unit (RPU). The omnidirectional aerials are connected to frequency receivers, also in the RPU. This unit processes the frequency and bearing content in each signal in binary digital form and passes it to the Operator's Console. The Situation Display in the centre of the Console shows all the signals which are being received, in one or more bands, against orthogonal frequency and true bearing axes. Further analysis of the signals by the Automatic Analyser can be initiated by the operator and the parameters displayed on the Alpha-Numeric Display immediately above the Situation Display. The Automatic Warner provides the facility of storing the parameters of some threat signals and automatically producing an alarm when any one of these signals is intercepted. The Manual Pulse Analyser, to the left of the Situation Display, enables the operator to examine a signal in more detail and to deal with new, unfamiliar types of signals.

To perform these functions, it is necessary that the specific characteristics of potentially hostile radars, frequency, pulse width, PRF, etc., which have been found by automatic or manual analysis, are stored in a section of the equipment generally known as the 'library'. Top-priority threat emissions, such as enemy missile radars, are stored in a section called the 'warner' for automatic identification and alarm; whenever emissions from such radars are intercepted in a real tactical situation an immediate alarm signal automatically warns of the presence of a priority threat. The identification of potentially hostile radar signals is carried out by automatic comparison of the parameters of the signal with the 'dictionary' stored in the 'library'. Fully automatic warning is given if any one pre-programmed hostile signal is detected. The surveillance function is carried out by displaying onto the operator's console screen all the emissions present in the air. The operator can quickly analyse and distinguish hostile emissions, and can initiate tracking in bearing of selected signals.

In the specific case of *Sheffield*, the Abbey Hill equipment did not perform either of these functions. No warning was given by the equipment, perhaps due to electromagnetic interference or perhaps because the radar parameters of the Exocet missile had not been stored in the 'warner', not having been programmed as a top priority threat.[5] The emission itself was not analysed since the operator did not have time to undertake the necessary operations.

The *Sheffield* was also equipped with two Corvus (or-Protean) chaff-launchers but these were not used for the simple reason that neither the missiles nor the aircraft they were launched from were discovered in time to feed the chaff-launcher with the necessary data. In order to be effective, 'chaff' a passive ECM, must be launched at precisely the right moment and in the precise direction and pattern to 'lure' the radar-seeker of the missile away from the ship.

Like most ships of the Royal Navy, the *Sheffield* was probably also equipped with active ECM devices: a Bexley 669 deception jammer for self-protection against missiles and a 667/668 noise jammer for jamming the search radars of hostile ships or aircraft, both of which had been originally designed to counter the Soviet *Styx* missile and other missiles of that generation. But also these devices were not activated for the same reasons given above.

Perhaps the most serious deficiency was the lack of an Infra-Red Warning Receiver (IRWR) and a more up-to-date deception jammer as these should have been the last line of defence for *Sheffield* when all its other systems had failed.

It must be pointed out, however, that devising ECMs capable of countering Western missiles like the Exocet, let alone the new generation Otomat, Harpoon and so on, is no easy task. The Exocet, although in use since 1973, incorporates several different types of sophisticated ECCMs which make it highly resistant to ECMs, including deception jamming. It is a missile of the 'fire and forget' type which means that, having launched the missile, the launching aircraft can immediately withdraw, thereby reducing the risk of being detected and shot down. The Agave radar[6] on board the aircraft only has to locate the target; once this has been done, data regarding the distance and direction of the target are automatically entered into the missile's computer-controlled guidance-system. All the pilot then has to do is launch the missile and go home; he does not even have to see the target. Once the Exocet missile is launched, it follows an initial course under inertial guidance, which is immune to ECMs, flying at an altitude of less than 30 feet above the sea under the control of its radar altimeter. At approximately 6 miles from the target, a small radar called Adac, which is located in the nose of the missile, turns itself on, acquires and locks on to the target and guides the missile to it. Adac is a monopulse tracking radar, operating in the X-band (8.5–12.5 GHz) and is highly resistant to ECMs. The monopulse technique is not new, having been used in Soviet SA-8 missiles and the more recent SA-10 and SA-11 missiles.

The Exocet also has other complicated anti-jamming and anti-deception devices. One such ECCM, known as frequency agility, enables the radar to change frequency when jammed. Another, called Home-on-Jam (HOJ) automatically homes the missile towards the source of the jamming; a third device, called Leading Edge, is highly sophisticated and top secret. Consequently, it is no easy task to build ECM equipment capable of deceiving or disturbing this type of missile.

However, this does not mean that such a task is impossible. What is certain is that the sudden outbreak of the Falklands war found some British ships, including *Sheffield*, lacking the latest electronic warfare equipment capable of countering technologically-advanced western missiles like the Exocet. The main reason, however, why ships like *Sheffield* were not adequately equipped is purely economic. Cuts in the British Defence budget had forced the Royal Navy to delay refitting the class to which *Sheffield* belonged. However, in spite of the cuts, the Royal Navy had to replace the Abbey Hill ESM systems with new Cutlass systems; old active ECM devices were also in the process of being replaced by Ramses 670 deception jammers and Millpost jammers.

It must also be pointed out that the first ECM devices installed on NATO warships had been built with Soviet anti-ship missiles in mind and would therefore probably be ineffective against a more sophisticated western missile. Moreover, the time element is of crucial importance where ECMs are concerned; they must be applied immediately and automatically, at the first sign of danger, which was not the case on board *Sheffield* where the Argentinian aircraft and missiles were sighted too late.

In the final analysis, it must be concluded that the Abbey Hill equipment on board *Sheffield*, if, indeed, it was in working order, was not able to distinguish and instantaneously interpret the electromagnetic signals coming from the radars of the Super Etendards and the Exocet missiles, either because of interference or because of its own intrinsic limitations.

On 7 May, Great Britain stepped up her naval blockade, declaring that all Argentinian military ships and aircraft encountered at a distance of over 12 miles from the coast of Argentina would be considered hostile and would be dealt with accordingly. A few days later, the Argentinian government announced similar restrictions for British ships and aircraft.

On 9 May, two British BAe Sea Harrier multi-role STOVL aircraft, flying a patrol, sighted the fishing-boat *Narwal* which had already been seen in the vicinity of the Task Force ships the previous week. Certain that it was an Argentinian spy-ship, the Sea Harriers dropped several bombs, one of which hit the ship, injuring fourteen men and seriously damaging the hull. The *Narwal* was forced to surrender and a Task Force helicopter arrived to pick-up prisoners. According to the British, the electronic equipment and documents found on board the fishing-boat, not to mention the presence of an Argentinian naval officer, provided clear indication that the ship was being operated for intelligence operations. On the same day, Task Force ships, supported by aircraft and helicopters, bombarded the Falklands for the first time, intending to disrupt Argentinian communication and Command and Control Centres.

The Russians probably furnished the Argentinians with data regarding the dispositions of the British Forces, collected by their numerous spy-satellites in orbits passing over the Falklands. Besides this source, the Argentinians also had four-jet Boeing 707 airliners modified for electronic surveillance and maritime reconnaissance,

Lockheed P-2V Neptune maritime patrol aircraft, Grumman S2F Trackers and Gates Learjet 35A aircraft, all built in the USA.

The British also appreciated the benefits to be derived from satellite surveillance and maritime reconnaissance systems. Although they did not have their own satellites, their ships were fitted with special Scot Skynet antennae which were able to receive data transmissions from the US Big Bird and the newer KH-11 satellites. The latter are generally considered to be the most sophisticated of all satellites, being able to receive and record earth images in digital form and immediately retransmit them to ground stations all over the world in a form which can be immediately utilised.

During the next few days the Russians with a special interest in EW and tactical operations sent further satellites into orbit which passed over the Falklands at twenty-minute intervals. One of these was the Cosmos 1372, for oceanic surveillance, equipped with nuclear-powered radar; the others were the Cosmos 1370 for photographic reconnaissance, the Molniya for communications and the Cosmos 1371 for SIGINT. A further small satellite for communications was also launched from the Salyut 7 space station which was already in orbit.

Meanwhile, the British Task Force was beginning to make preparations for a beach-landing in the Falklands. The British increased air strikes and naval bombardments of Argentinian coastal military installations and were carrying out the following pre-landing actions:

clandestine reconnaissance of the islands in order to choose a suitable beach

clearing the selected landing site of all natural and man-made obstacles on the sea-bed by special underwater demolition teams

installation on East Falkland island of special automatic electronic sensors to provide data regarding the deployment and movement of Argentinian troops on the island

commando raids on the various islands to destroy stores and installations (the raid on Pebble Island was particularly successful, the British destroying ten Argentinian Pucara aircraft and a large ammunition dump)

diversive actions on beaches other than that chosen to confuse the Argentinians as to where the landing would in fact take place.

Use of Chaff by a Warship Against an Anti-Ship Missile

Chaff was used extensively in the Falklands conflict. Modern chaff is manufacturerd from a variety of materials, including aluminium-coated glass filament (fine dipoles having a low fall rate), cut aluminium foil and silver-coated nylon filament.

At the same time, both the British and the Argentinians were building-up their forces in preparation for the final battle. Britain dispatched six more warships, twenty more Harriers and the luxury liner the 67,140 ton *Queen Elizabeth 2*, carrying 3000 soldiers. Having been made aware of the electronic shortcomings of some of their ships in the *Sheffield* incident, large quantities of chaff were also dispatched for use by such ships during air attack. Tactics were also studied for using helicopters to launch chaff. Subsequently, chaff was frequently used to blank out enemy search radar or to divert approaching enemy missiles. For further protection from attacking aircraft, the British devised a method of launching chaff from ships' funnels, mixed in with the exhaust gases. However, the employment of chaff in the South

Atlantic did not always achieve the desired result as it was often dispersed by the gale force winds.

The Argentinian Expeditionary Force, consisting of roughly 10,000 men was equipped with German-built wire-guided anti-tank Cobra 2000 missiles, night-vision devices, the up-to-date Franco–German Roland SAM, and FMA IA-58 Pucara and Aermacchi MB 326G and MB 339 ground-support aircraft.

On 21 May, two hours before sunrise, the British landing operation to regain the Falklands got underway: Task Force ships sailed into San Carlos Water and began shelling coastal batteries near Port San Carlos. This was followed by the landing of 2500 men, mainly Royal Marines and Paratroops, who established a beach-head in San Carlos Bay, well sheltered from the South Atlantic gales. The Argentinians, who had not expected the landing to take place at Port San Carlos, did not put up much resistance. Their retaliation arrived, however, from the air, in the form of MB 326s, Skyhawks and Mirages that furiously bombed and rocketted the British ships in the Bay, five of which were hit. One of them, *Ardent*, a 3250-ton Type 21 frigate, was very severely damaged and, again, caught fire violently; twenty-two men died, thirty were injured and the ship continued to burn uncontrollably until she finally sank.

On 22 May, the British beach-head was consolidated by landing a further 2500 soldiers at San Carlos; they were equipped with night-vision devices (light intensifiers and infrared goggles), Scorpion light tanks, armoured vehicles with Rapier SAMs, Blowpipe man-portable SAMs, 105 mm air portable light guns, mortars and several anti-aircraft radars.

On 23, 24 and 25 May, the Argentinians made a series of air attacks on the British beach-head; wave after wave of Skyhawks and Aermacchis, supported by Mirage and Dagger fighters repeatedly bombed both the beach-head and the Task Force ships in the Bay of San Carlos. On 23 May, during one of these attacks, the British Type 21 frigate *Antelope*, on a reconnaissance mission in Falklands Sound was hit by a 5000-lb bomb which penetrated the engine room, but did not detonate. It exploded while bomb experts were trying to de-fuse it, killing two officers, and breaking *Antelope*'s back.

In spite of extremely heavy losses, the Argentinian aircraft continued their day-long attacks on the British ships with great courage and skill throughout 24 and 25 May. At 18.30 on the 25 May, a formation of Skyhawks bombed and sank the Type 42 destroyer HMS *Coventry*. Another air formation including Super Etendards armed

Table 1. Naval Attrition

Ship	Displacement (tons)	Attacking Platform	Weapons
Royal Navy: Sunk			
HMS *Antelope*	3250	Mirage/Skyhawk	Bombs
HMS *Ardent*	3250	MB.339	Rockets
HMS *Coventry*	4100	Skyhawk	Bombs/Rockets
HMS *Sheffield*	4100	Super Etendard	AM.39 Exocet
HMS *Sir Galahad*	5674	Mirage/Skyhawk	Bombs/Rockets
Merchant Navy: Sunk			
Atlantic Conveyor	14,950	Super Etendard	AM.39 Exocet
Royal Navy: Damaged			
HMS *Antrim*	6200	MB.339/Skyhawk	Bombs/Rockets
HMS *Brilliant*	4000	MB.339/Skyhawk	Bombs/Rockets
HMS *Broadsword*	4000	MB.339/Skyhawk	Bombs/Rockets
HMS *Glamorgan*	6200	Coastal Site	MB.38 Exocet
HMS *Glasgow*	4100	MB.339/Skyhawk	Bomb
HMS *Plymouth*	2800	Mirage/Skyhawk	Bombs/Rockets
HMS *Sir Tristram*	5674	Mirage/Skyhawk	Bombs/Rockets
Armada Republica Argentina: Sunk			
ARA *General Belgrano*	13,645	HMS Conqueror	Mk 8 Torpedo
ARA *Santa Fe*	2420	Sea King	Gunfire
Armada Republica Argentina: Damaged			
ARA *Alferez Sobral*	850	Lynx	Sea Skua ASM

with Exocet missiles, headed for a large target which they mistook for the aircraft carrier *Hermes*; it was in fact the container-vessel *Atlantic Conveyor*. The ship, which was transporting Wessex and Chinook helicopters and spare parts, was hit by an Exocet. She was badly damaged and sank shortly after the crew had abandoned ship. The loss of equipment on board the *Atlantic Conveyor* was a very severe blow to the Task Force. The tactics used in this attack were almost identical to those employed against *Sheffield*. When the pilots of the Super Etendards 'popped-up' to an altitude of 500 feet to check the situation in the area, their radar screens showed a large target surrounded by several smaller targets, which were the escort ships. As soon as they were alerted, the escort ships began to launch huge quantities of chaff which was effective in confusing and deviating the Exocet missiles. However, quite by chance, one of the wandering missiles hit the *Atlantic Conveyor* which, being a merchant ship, had no electronic self-protection equipment.

During the next few days, the 6200-ton 'County' class destroyer *Antrim*, the 4000-ton Type 22 destroyer *Broadsword* and the 3200-ton Exocet-armed 'Leander'-class frigate *Argonaut*, as well as several landing-craft and logistic vessels, were damaged during Argentinian air raids. During these attacks, the Sea Wolf anti-missile system installed on *Broadsword* was used for the first time, scoring a hit and destroying an Argentinian Skyhawk. Another Task Force ship, *Brilliant*, was also equipped with the Sea Wolf system, but neither ship had an opportunity to use these anti-missile missiles against the Exocet during the Falklands war.

The tactic adopted by the Argentinian pilots was simple but clever. Their attacks were carried out at dusk in formations of from four to ten aircraft of various types. These would all head for the same target in order to 'saturate' the ship's radar and other AA defences. They flew in almost at sea-level and headed for the northern tip of the archipelago, using the islands and their hills to shield them from the British ships' radars. They would then suddenly swing round and all appear simultaneously from behind the coastal ridge of the northernmost island and attack their chosen target-ship from all sides. The British radar operators were unable to track all the hostile aircraft simultaneously and one or two of them almost always managed to slip through and launch their bombs and rockets. The ships' ESM equipment also proved ineffective, when this saturation tactic was used, as the Argentinians flew with their radars switched off, so there were no electromagnetic emissions in the air to be picked up.

Table 2. Missile Systems

Missile	Maximum Range miles (km)	Guidance System
Surface-to-Air		
Blowpipe	1.9 (3)	Optical
BAe Rapier	4.6 (7.5)	Optical-Beam Rider
Roland	3.9 (6.3)	Semi-Active/IR
Sea Cat	2.9 (4.7)	Radio Control
Sea Dart	24 (38.6)	Semi-Active
Sea Wolf GWS-25	1.9 (3)	Semi-Active
Tiger Cat	2.9 (4.7)	TV-Optical
Air-to-Air		
AIM-9L Sidewinder	11 (17.7)	Passive IR
Air-to-Surface		
AM.39 Exocet	31–43.5 (50–70)	Active Radar
Sea Skua	9.3+ (15+)	Semi-Active

Meanwhile, the troops which had landed at Port San Carlos, by this time well-organised from a logistic point of view, began their march towards Port Stanley, following two routes. One group marched towards Douglas and Teal Inlet over very difficult terrain while the other group marched towards Darwin and Goose Green in the southern part of the island.

On 27 May, a major battle for the conquest of Goose Green airfield began. The Argentinian troops were supported by Pucaras and the British troops by Harriers. The battle lasted about fourteen hours, most of it fought by night, to the advantage of the British troops who were equipped with light intensifiers and infrared goggles and were thus able to employ NATO night combat tactics.

The Argentinians put up strong resistance but were unable to prevent the two important locations of Darwin and Goose Green falling into British hands. It was soon apparent that the British soldiers, all volunteers, were far superior to the Argentinians, most of whom were very young, inexpert conscripts. Moreover, the cold climate of the Falklands favoured the British Marines and Paras who had been acclimatised to cold environments during NATO training exercises in Northern Europe beyond the Arctic Circle.

Both sides suffered heavy losses in the battle of Goose Green. According to the British, the Argentinians lost 250 men, while 1400 were taken prisoner. The British suffered seventeen dead and thirteen wounded. The conquest of Goose Green provided the British with a base from which they would advance towards Port Stanley.

30 May brought yet another fierce Argentinian attack on the Task Force which had meanwhile stepped up its air and naval bombardments of installations at Port Stanley. The Task Force was stationed at a distance of 95 miles to the north-east of the Falklands, from where Sea Harriers took off from *Hermes* and *Invincible* to attack Port Stanley. Taking part in the Argentinian attack were two Super Etendards, one carrying the last remaining Argentinian Exocet missile, four A-4 Skyhawks and six Mirages and Daggers (Israeli-built Mirage developments) which had the task of distracting and busying the radars on board the British ships. Before the Super Etendards came in to attack, Skyhawks and Daggers, approaching from the east, managed to get round British ground anti-aircraft defences and attract the attention of the British radars and lure up the interceptors from the *Invincible*. While this was going on, the Super Etendards came in to launch the last remaining Exocet missile which, according to Argentinian sources, hit *Invincible*. Two Skyhawks were shot down during the attack, and the British repeatedly denied any damage to *Invincible*.

The Argentinians lost about a third of their aircraft in the course of these air attacks. The lack of equipment for electronic warfare on most of their aircraft no doubt contributed greatly to these heavy losses. The only aircraft which had such equipment were the Super Etendards and Daggers which were fitted with RWRs, provided by the French and the Israelis, respectively.

British losses, on the other hand, were aggravated by their choice of Port San Carlos as the site of their landing-operation. The effectiveness of the British air defence radars was considerably reduced by the clutter created by the surrounding hills. It is also worth pointing out that British losses would have been much greater if all the bombs that hit their ships had exploded. The failure of many bombs to explode was probably due to the fact that the Argentinian pilots were forced to fly so low that there was not enough time between launching and impact for the bombs to auto-activate.

The British troops continued their advance towards Port Stanley using a 'leapfrog' tactic which involved covering short distances rapidly by night. Attacks were preceded by air and naval bombardments of Argentinian defences and were supported by artillery and mortar fire aided by infrared and electro-optical aiming systems.

Firing was directed and coordinated by three interacting electronic systems. The first, called FACE was a mini-computer which calculated firing data; the second, called ALICE, automatically transmitted this data to field artillery; the third, called AWDATS, programmed the simultaneous firing of twenty-four artillery pieces at various locations.

British mortar and gunfire was extremely accurate, as a result of these electronic systems, and Argentinian positions suffered heavy punishment and their radars and other communications systems were frequently put out of action. The British also had an excellent information service based on interception of tactical communications and on reconnaissance carried out by special scouting squadrons. In this way, British commanders always knew where the enemy was and what he was doing. On one occasion, just after a bombardment, they intercepted a radio message in which Brigadier General Mario Benyamin Menendez expressed his fear that, if things went on that way, the Argentinian situation might worsen rapidly; the British, therefore, had the advantage of knowing just how precarious the Argentinian situation was.

On 6 and 7 June, with the final attack on Port Stanley close at hand, numerous commando raiding parties of highly-trained men were sent behind Argentinian lines to destroy radar and radio installations, with the aim of paralysing enemy communications. Again in the field of communications, the British misinformed the enemy by means of deceptive measures, propaganda and infiltration, greatly assisted by the cooperation of the Falkland islanders.

Besides having their messages read by the British, the Argentinians also had the misfortune to receive information from the Russians which often turned out to be mistaken or out-of-date. On the other hand, the Argentinians could always get accurate information regarding British actions via interception of tactical communications between aircraft, ships and ground forces.

On 8 June, the Argentinian air force initiated another series of deadly attacks on British ships and troops in the Port Stanley area, causing the British to delay their assault on the capital. During one of these attacks, the two landing-ships, *Sir Tristan* and *Sir Galahad*, were badly hit by Argentinian aircraft, causing many casualties among the troops who were trying to land. The Argentinians were greatly helped by a mobile radar unit, a Westinghouse AN/TPS-43, which had been set up in a place called 'Sapper Hill'. This large American 3D radar was part of the Command Information and Control Centre (CIC) which the Argentinian air force had set up in the Falklands to coordinate air defence operations. The British made several attacks on this radar, one using anti-radar AGM-45 Shrike missiles launched by a Vulcan long-range bomber. However, these attacks were unsuccessful and the radar functioned effectively right up to the end of the war.

On 11 June, during another Argentinian air attack in the Falklands Channel, the 2800-ton frigate *Plymouth* and the assault-ship HMS *Fearless* were badly damaged.

Meanwhile, British troops were getting closer to Port Stanley and on the night of 11 June, with the help of Lynx helicopters armed with rockets, they made a surprise attack on enemy defences. The Argentinian soldiers were sleeping but nevertheless reacted with fierce determination, engaging in hand-to-hand combat for several hours. They were finally forced to retreat, however, leaving Two Sisters hill to the British.

Meanwhile, Task Force ships continued to bombard installations in the area of Port Stanley. On 11 June, during one of these shore bombardments, the large 6200-ton missile destroyer *Glamorgan* was hit by an MM-38 Exocet missile launched from a shore-battery; the ship was hit in the stern, 2 metres above the water-line and, although the missile did not explode, ten members of her crew were killed and seventeen injured. *Glamorgan* had been located by the Westinghouse AN/TPS-43F radar and, given the accuracy of this sytem and that of the Exocet missile itself, it seems strange that the ship was hit only at an extremity and not mid-ship, as in the case of *Sheffield*, especially since she was drawing near the coast at the time. Unofficial British sources attributed this to the combined use of active ECMs (presumably a Bexley deception jammer) and chaff.

The British had by now occupied all the hills surrounding Port Stanley and, on 12 and 13 June, kept up a steady, accurate and selective bombardment on the Argentinian Garrison who had withdrawn into the built-up area of Port Stanley itself. In these last phases of the air-to-ground battle, laser beams were used as support measures for aircraft at close quarters. In particular, laser-guided bombs were launched by Harriers against laser-illuminated targets, employing tactics similar to those used by the Americans in Vietnam.

The first attacks took place in the area between Two Sisters farm and Mount Tumbledown. Two separate attacks were made using Harrier GR3s which launched Paveway laser-guided bombs[7] from a distance of 6–7 kms. In both cases, the bombs missed their target as the laser was switched on prematurely. Both times the Harriers, each carrying two bombs, came in from the south-west at an altitude of 500 feet, their approach masked by Mount Harriet. The pilot, guided by a pre-determined landmark, would then drop altitude and, at a pre-programmed waypoint, launch the bombs, informing the FAC (Forward Air Controller) at the moment of launch. He would then make his escape, never having sighted the target optically. The FAC, from a vantage-point which enabled him to see the target without himself being seen by the enemy, would then, shortly after receiving the communication, direct the laser-beam onto the chosen target. This system was also used to bomb the airfield at Port Stanley but the results were rather disappointing.

Special weapons were also used in an attack on the above-mentioned Westinghouse AN/TPS-43 radar used by the Argentinian air force. At the beginning of the war, two Vulcan B2 long-range bombers had been equipped for radar suppression tasks with four anti-radiation AGM-45 Shrike missiles. They operated from Wideawake airfield on Ascension island, the nearest airbase to the Falklands available to the British. The flight to the Falklands involved complex pre-planning and several in-flight refuellings from Victor tanker aircraft. The Vulcan raids on the Falklands were the longest combat missions ever flown. One Vulcan, serial XM 597, was used to carry out the first mission on the night of 28 May. This mission was never completed, however, due to difficulties with the in-flight refuelling from one Victor tanker. Two days later, the mission was repeated, with the attack by the Vulcan being coordinated with a Task Force Harrier strike, but its outcome was uncertain. The outcome of the last attack, made on 2 June, was inconclusive; the same Vulcan, XM 597, stayed in the area for almost an hour trying to provoke enemy radar emissions but the Argentinians switched off the radar whenever the aircraft approached for attack and so the Shrike missile was denied the emissions it needed to guide itself to the radar.

However, by this time the British had the Argentinian garrison firmly in their grip and, after another surprise attack on 12 June, the latter had no alternative but to request an armistice, thus putting an end to the war.

At the end of the war, both sides issued reports on losses suffered and those inflicted on the enemy but these figures did not tally. The British maintained that only one Harrier and no Sea Harriers had been shot down by missiles.[8] This type of aircraft took part in many battles in the Falklands and the fact that it avoided being shot down by missiles is probably due to the superiority of British electronic warfare equipment, all British aircraft being equipped with RWRs, chaff-launchers and IR flares. Harrier pilots declared that they had often managed to avoid Argentinian Roland SAMs by making an abrupt evasive manoeuvre as soon as their RWR warned them that a missile was on its way. Although Sea Harriers can carry jammer pods, the only British aircraft integrally equipped with active ECM devices, however, were Vulcan bombers. Prior to their departure for the Falklands, these aircraft were equipped with US AN/ALQ-101 jammers housed in pods which had been taken from BAe Buccaneers. This is no doubt one of the reasons why no Vulcans were lost.[9]

The Royal Navy Sea Harrier, being a multi-role aircraft, was already equipped to carry the US AIM-9L Sidewinder air-to-air missile, whilst the RAF Harrier tactical attack aircraft was hastily modified to carry AIM-9Ls AAMs as the Task Force sailed for the Falklands. the AIM-9L belongs to the third generation of the famous infrared guided AIM-9 Sidewinder family which is in service with many NATO and other western air forces. Being designed with an 'all-aspect attack capability' (ALASCA), it has radically altered the air combat equation since, unlike previous IR missiles, they can also be launched head-on. During the Falklands conflict twenty-four out of twenty-seven AIM-9L missiles launched successfully brought down Argentinian planes, a record which speaks for itself.

Assessing the Falklands war from the point of view of electronic warfare, several innovations were employed in ground combat, such as extensive use of night-vision systems and new night-combat tactics. However, nothing really new emerged in the air and naval fields. Argentinian use of radar in air combat was rather limited and this, together with the fact that British forces were equipped for electronic warfare in situations quite different from those in which they actually had to fight, meant that the real capabilities of electronic warfare systems could not be exploited to the full. A major factor is training: the British captured considerable quantities of Argentinian night-fighting and similar sophisticated electronic equipment, such as IR-goggles, and, if anything, the individual Argentinian was better equipped but lacked the training and motivation of the British professionals. It should also be mentioned that some ships of the Royal

Navy were poorly equipped to deal with new threats, such as the western anti-ship missiles, in such a confused electromagnetic environment. The Argentinians made very little use of electronic warfare systems in air attacks. On the other hand, they distinguished themselves in the field of passive electronic warfare, using rapidly-modified airliners, such as the Boeing 707, for ELINT missions and ESM.

22

The Lebanon

At 11.25 on 6 June 1982, after two days of heavy air attacks and naval bombardments, the Israeli armed forces launched their long awaited and greatly feared attack on Palestinian strongholds in south Lebanon. Their declared aim was to create a 50 km wide buffer zone along the Israeli–Lebanese border to prevent the Palestinians from attacking Israel.

As Israeli troops advanced northwards, easily overcoming Feddayeen resistance, the danger of a confrontation with the Syrian ADF (Arab Dissuasion Force) stationed in the Lebanon became more and more likely. The situation exploded on Thursday, 9 June as the Israelis approached the Bekaa Valley where 600 Syrian tanks were based, protected by an AA 'umbrella' consisting of twenty surface-to-air missile batteries: Russian-made SAM-6 (mobile) and SAM-2 and SAM-3 (fixed)

The entire Syrian air force was on the alert. When, in the afternoon, the three central columns of the Israeli armoured forces came into contact with the advanced guard of two Syrian armoured brigades (see illustration page 266), the Syrian high command immediately sent sixty MiG-21s and MiG-23s to provide close air support for their tanks. The Israelis were not taken by surprise, however, as their US made E-2C Hawkeyes, with their enormous Early Warning radars, were already orbitting along the Lebanese coast to watch for the take off of any Syrian aircraft from their bases inside Syria and guide the Israeli fighters to them. They thus immediately sent out a total of ninety aircraft: ultra-modern US-built F-15 McDonnell Douglas Eagle and General Dynamics F-16 Fighting Falcon fighters for air-combat, Israeli-made IAI Kfirs and veteran McDonnell Douglas F-4 Phantoms for air-to-surface attacks and McDonnell Douglas A-4 Skyhawks for close air support. A four-jet Boeing 707, equipped for EW, was also sent up to intercept and jam enemy radars and communications from outside the range of Syria's weapons (Stand-off jamming). As the Syrian aircraft approached the 'hot' zone, radio communications between them and ground commands were jammed by the Israelis, thus cutting off route and attack instructions.

The Israeli pilots, on the other hand, were perfectly vectored by the E-2Cs to the optimum positions from which to attack the Syrian MiG

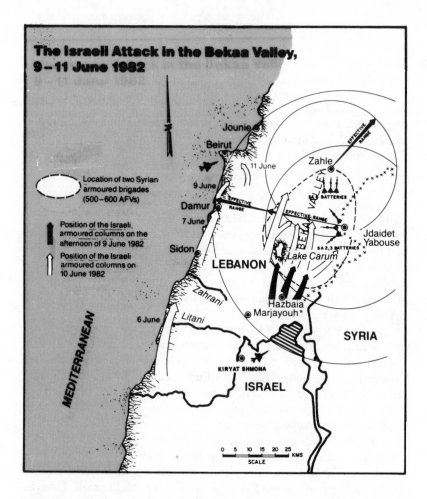

The Israeli Attack in the Bekaa Valley, 9–11 June 1982

aircraft. The Israeli aircraft were all equipped with the latest automatic computer-controlled EW equipment, as well as laser devices for target designation, Infrared AIM-9L Sidewinder air-to-air missiles and AGM-45 Shrike and AGM-65 Maverick anti-radar missiles. Thus, confident in their utter professionalism and electronic superiority, the Israeli pilots flew at full speed towards the enemy.

Every Israeli aircraft was equipped with an HUD (Head-Up Display) which greatly reduced the pilot's workload. In this system, data for navigation and combat are calculated by computer and transmitted to a presentation processing unit which transforms the data into blue and orange phosphorescent optical images which are then projected onto a glass screen just behind the windscreen. HUD systems usually operate in conjunction with a radar or LLLTV system and provide the pilot, whatever the conditions of visibility, with an accurate 'picture' of his immediate surroundings and the enemy air situation so that he is not distracted by constantly having to look down at the various flight instruments in the cockpit or make difficult navigational calculations.

The Israeli aircraft were also equipped with the very latest fully automatic, computerised deception jammers which were able to send even the most advanced missiles off course, thus ensuring the survival of the pilot, and RWRs which immediately warned him that his aircraft was being 'locked-on' by a tracking radar or the radar-seeker of a missile itself. Each Israeli aircraft was also equipped with expendable passive countermeasures, both chaff and IR flares, to be launched at the right moment to 'distract' oncoming missiles.

As soon as the ninety Israeli aircraft penetrated the air space over the Bekaa valley, they were immersed in a huge mass of electromagnetic emissions coming from hundreds of enemy radars and radios present in the zone. In moments such as this, it is imperative for the pilot to immediately analyse and identify all air defence radars and ground-to-air missiles and to determine the position of enemy interceptors, all of which constitute a serious threat to his survival. Neither his own brain nor traditional avionics are able to deal with such a large number of threatening signals and ascertain which constitute the greatest dangers. It is at this point that computer 'software' (i.e. all the programmed information and logic previously and preventively fed to the computer) proves to be the vital asset. In this way, the Israeli pilots were able to approach the enemy aircraft, mainly by following the vectors received from the E2-C Hawkeyes.

The RWR on board his plane would alert the pilot that the radar of a SAM battery had locked on to his aircraft. Almost instantaneously, the EW computer would analyse and identify the various threats, determining their priority and indicating the most effective defensive action to counter each individual threat.

The battle went on from 9 to 11 June. Throughout the combat, the Israelis made extensive use of deception jammers to divert electronically-guided missiles and IR flares to divert IR-guided missiles. As soon as an Israeli pilot located a Syrian MiG via his HUD, all he had to do was to superimpose the firing symbol on his HUD over the enemy aircraft, push the relevant button to operate the most appropriate arms system decided by the computer. All the real work was done by the IR sensor of the implacable Sidewinder.

The Syrian aircraft, on the other hand, were not furnished with EW equipment, as the Russians usually remove such equipment from aircraft supplied to foreign countries. The Syrian pilots were also at a great disadvantage because their radars and radio communications were jammed by the Israeli Boeing 707. Furthermore, support from ground AA batteries was very limited due partly to Israeli jamming of their radars and partly to the large number of aircraft present in the sky which meant that there was a great risk of hitting their own planes.

As usual, there are discrepancies between declared losses of both sides. The Israelis affirmed that they had shot down seventy-nine enemy planes, damaged at least seven and destroyed nineteen of the twenty deployed SAM-2, SAM-3 and SAM-6 batteries while their own sustained losses amounted to only one aircraft. The Syrians declared that they had shot down nineteen enemy aircraft.

Besides their extensive use of EW systems, the Israeli victory must be in part attributed also to new tactics which were used for the first time in this battle. The most important of these was the tactic used for attacking SA-6 missile batteries. Some time previously, the Israelis had installed a certain number of mock-up SA-6 batteries in the Negev desert for training purposes, against which they flew both ordinary aircraft and RPVs.

The suppression of enemy AA defences is an indispensable preliminary to actions which require the penetration of enemy air space and the establishment of air supremacy. The Israeli Air Force, therefore, set about destroying, both before and during the air battle in the skies of the Lebanon, Syrian ground-to-air missile batteries which constituted a deadly threat to their own strike aircraft. All available means were used to this end, including the Israeli-built Scout and Mastiff RPVs.

These RPVs are very small—with a wingspan of just 3.60 m, the Scout is only 3.51 m long and 0.94 m high; moreover, they are made of fibreglass, which is transparent to radar. Consequently, they are difficult to detect and locate by enemy radars and are thus able to penetrate enemy air space with minimal risk of being shot down. For this reason, they are ideally suited to the tasks of battlefield reconnaissance and surveillance. Some versions were equipped for such missions with a TV camera with zoom lenses and a transmitting system which could send back to it's ground controller a continuous flow of pictures of enemy positions. Other versions were equipped with a radar reflector which returned radar echoes comparable to those of an attack aircraft. Others functioned as ESM platforms, intercepting and analysing enemy radar emissions and retransmitting them to ground stations or to an aircraft in flight. Finally, some were equipped as laser-designators to illuminate a target to be attacked by laser-guided missiles.

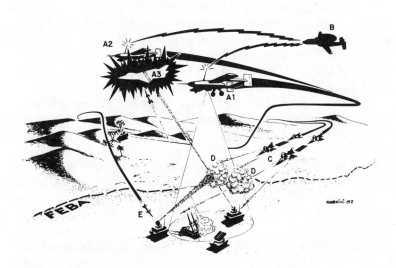

Israeli SAM-6-Suppression Tactic

An artist's impression of a typical Israeli air force air strike to seek and destroy Syrian SAM sites. One RPV, **Al**, carried a television camera which provided visual reconnaissance of SAM sites; the other RPV, **A2**, beamed signals which deceived the Syrians into believing that the tiny fibreglass aircraft were Israeli jets, **A3**. The Syrians turned on their missile-associated radar systems, allowing the RPVs to 'fingerprint' the radar emissions and relay this information to the E-2C command and control aircraft which vectored Israeli strike aircraft to attack the SAM sites, **B**. The destruction of the missile's radars was accomplished by both air-to-surface ARMs launched by F-4 Phantoms, **C**, and surface-to surface ARMs, **E**. The attackers defended themselves by launching chaff, **D**, and IR flares to prevent lock-on by tracking radars and missile's homing IR seekers. Without the radar, the SAM launchers were blind and were destroyed by cluster bombs dropped by F-15s and F-16s.

The anti-SAM operation (see page 269) began with a series of reconnaissance flights made by RPVs equipped with TV cameras. As soon as one of them discovered a SAM battery and transmitted the images to ground command, two more RPVs were sent out, one, the radar decoy, would simulate an attacking aircraft with the intention of provoking the SAM battery into switching on it's radar, while the other, the ESM craft, intercepted emissions from the SAM radar, analysed them and retransmitted them to airborne E-2Cs and Boeing 707s. The emissions received were then processed by the computers on board these aircraft to produce, in real time, data for the guidance of anti-radar missiles. One of these aircraft then gave the order for the launching of an Israeli-built Zeev surface-to-surface anti-radar missile, if the SAM battery was within its range of about 40 kms, otherwise, a Shrike-armed F-4 Phantom strike would be ordered.

Sometimes, the Syrians realised that the enemy was using RPVs to intercept the emissions of their radars and thus launch anti-radiation missiles at them and would immediately turn off the radar in question to deprive the Israeli missiles of their electromagnetic homing beam. In this case, the Israelis would send out a laser-designator RPV and an attack aircraft armed with laser-homing AGM-65 Maverick missiles. Once the radar had been destroyed, the now blind SAM battery was attacked by cluster bombs which destroyed both the missiles and their associated vehicles.

By this alternate use of RPVs and Phantoms, perfectly coordinated by the E-2C and Boeing 707 aircraft, the Israelis managed to destroy nearly all SAM batteries in the area, thus depriving the Syrian armoured columns of AA defence.

The unprecedented successes of these air operations were no doubt enhanced by the experience gained during the war between Iraq and Iran when the Israelis made a lightning air attack on the Iraqi nuclear reactor under construction at Tammuz, about 20 km from Baghdad. The raid was carried out on 9 June 1981 by a formation of eight F-16s and F-15s flying at very low level. They flew over the northern part of Saudi Arabia, along the Jordanian border and through Iraqi air space as far as Tammuz. They attacked the target and, without meeting any resistance, took the direct route through Jordan back to Israel. As a result of the meticulous planning and use of ECMs, the Israeli aircraft managed to avoid detection by Saudi, Iraqi and Jordanian radar surveillance as well as the very advanced airborne radars of the AWACS on loan to Saudi Arabia from the US Air Force!

Another new air combat tactic used for the first time in the Lebanon by Israeli pilots was that of effecting the final phase of an attack against enemy aircraft from the side in order to have a larger target area.

On the ground, outstanding results were achieved by laser rangefinders, laser target-designation systems and wire-guided TOW missiles, all controlled by computer. During one of the first actions between Israeli armoured columns and the advanced elements of the two Syrian armoured brigades, which took place on 9 June at the southern end of the Bekaa valley, the Syrians lost about sixty T-55 and T-62 tanks. The next day, further north in the Bekaa valley, there was a major battle between 300 to 400 Israeli tanks and about the same number of Syrian T-55, T-62 and T-72 tanks deployed along the road running from Beirut to Damascus. Israeli artillery and helicopters armed with TOW missiles also took part in the battle and, according to the Israelis, all the Syrian tanks were put out of action, many of them being captured.

Alarmed by the poor performance of the aircraft and missiles they had supplied to Syria, the Soviets immediately sent a group of experts, headed by the vice commander of the Soviet air force, to the Lebanon and had a damaged T-72 tank sent to Russia for examination.

As a first step towards making up for the losses of the SAM-6 launch-pads, the Russians supplied the Syrian air force with a number of ground-to-air SAM-8 *Gecko* and SAM-9 *Gaskin* missile systems. The SAM-9 was first seen in November 1975 in the annual military parade to commemorate the October Revolution. It was, therefore, by no means new, but, in Lebanon, it almost immediately succeeded in shooting down an F-4 Phantom on 25 July, probably because the Phantom was not electronically prepared to meet the new IR threat. The Israelis later managed to capture SA-8 and SA-9 missile systems, as well as T-72 tanks, about which systems little was known in the West. This led to the development of ECMs which enabled the Israelis, in September 1982, to destroy five or six such missile batteries in the Lebanon.

Generally speaking, these battles in the Bekaa valley proved the effectiveness of the coordinated use of electronically controlled arms with the necessary EW back-up. But, above all, those air and surface conflicts together provide the first example of warfare in 'real time'; warfare in which air reconnaissance, the distribution of its results to attacking forces and the attacks themselves were carried out almost simultaneously in rapid succession, closely coordinated with the extensive use of EW systems. The outstanding results achieved by the Israelis show that the new concept of 'real-time' warfare, supported by accurate planning of EW actions, was the real key to their success. Military commanders from all countries should give thought to the events which took place in the Bekaa valley in July 1982 as they give an idea of what future battles will be like.

French and US air actions over the Lebanon

The Israeli victory in the Bekaa valley in 1982, like their previous victories in 1967 and 1973, did not solve any of the problems of the Middle East. The bloody war in the Lebanon dragged on inexorably. Throughout 1983, there was a long sequence of battles between the opposing factions, assaults, massacres, air raids, Israeli aircraft and RPVs shot down, all culminating in the terrible truck-bomb attacks on the barracks of French soldiers and US Marines of the international peace-keeping force in October 1983, in which hundreds of men were killed and injured.

Of course, it was not long before reprisals were sought and the first to take action were the French. On the afternoon of Thursday, 17 November 1983, eight Super Etendards took off from the aircraft carrier *Clemenceau*, which was cruising about 100 miles off the Lebanese coast. The aircraft headed inland towards Baalbek to attack the old Lebanese barracks of Cheik Abdallah, occupied by Islamic volunteers suspected of having organised the bloody attack on the barracks of the French paratroopers.

The target-area assigned to the pilots of the Super Etendards was very small and a thorough air-photographic reconnaissance had been carried out a few days previously. Orders for the mission had come from Paris but the choice of the day had been left to Rear-Admiral Klotz on board the *Clemenceau*, depending on local weather conditions and other relevant factors.

The eight Super Etendards each carried one 400-kg bomb and three 250-kg bombs. Air cover was provided by two French F-8E (FN) Crusader fighters flying patrol above the Super Etendards, ready to intervene if Syrian aircraft should appear. However, the chances of success of such a mission would have been very slim indeed without the protection of stand-off EW-equipped aircraft.

It is no secret that the Syrians, through Russian aid, have set up unprecedented AA defence coverage of Lebanese airspace comprising: SAM-2 missiles with a range of 25–30 miles, SAM-3s with a range of 16–19 miles, SAM-5 (based in Syria) with a range of 185 miles, SAM-6 with a range of 19–38 miles and SAM-8 with a range of 8 miles (all with radar guidance), portable IR-guided SAM-7 missiles with a range of 6 miles and new IR/radar-guided SAM-9 missiles with a range of 5 miles and numerous conventional AAA batteries of the 23 mm ZSU-23-4 and 57 mm ZSU-57-2 types (both with radar guidance).

The need for EW-equipped aircraft was obvious and, since the *Clemenceau* was not carrying any such aircraft, the French sought help from the US Sixth Fleet which had for some time been present in the waters of the east Mediterranean. The Americans provided Grumman EA-6B Prowlers for stand-off jamming; during the entire raid carried out by the Super Etendards, the Prowlers jammed enemy search radar and weapons-guidance radar from a position outside their range.

Guided by their inertial navigation systems (INS), the French pilots managed to achieve a surprise attack, each group making a single-pass attack of first two, then four, then two aircraft. The first six dropped their bombs as scheduled but the last two were prevented from doing so by fierce AA fire from the ground.

On completion of the mission, the French air commander decided that, as the massive Syrian air defences were now alerted, it would be unwise to send a photo-reconnaissance Etendard IVP to confirm the results of the mission. Nevertheless, the mission was held to have been technically 100 per cent successful.

A few weeks later, on 4 December 1983, the Americans organised a reprisal mission against Syrian missile bases which had, in the last few days, attacked US reconnaissance aircraft patrolling the area. At dawn, sixteen Grumann A-6E Intruders and twelve LTV A-7E Corsair II attack aircraft took off from the aircraft carriers *John F. Kennedy* and *Independence*. As is usual for this type of mission, called air defence suppression, E-2C Hawkeye Early Warning aircraft were first sent on missions of stand-off deep search to locate Syrian missile batteries. A number of CAP (Combat Air Patrol) Grumman F-14 Tomcats escorted the formation from a distance. Last, but not least, there were, of course, the ECM-equipped EA-6B Prowlers which had the task of jamming enemy radars.

The task of suppressing the Syrian AA batteries was shared by the A-7E Corsairs and the A-6E Intruders: the former had to attack the SAM radars and AAA radars with Shrike anti-radiation radar missiles (ARM) while the latter destroyed the missile launchers.

However, the departure of so many aircraft did not escape the notice of the Soviet cruiser and spy-ship which, as usual, were shadowing the US Sixth Fleet formation. Presumably, fearing another disaster in which the weapons they had delivered to the Syrians might be demolished and the Soviet military advisors in the zone harmed, the Russians immediately informed Damascus via radio of the departure of the large US air formation. Consequently, the Syrians had all the time they needed to prepare a suitable reception for the pilots of the US Sixth Fleet.

Thus the US aircraft encountered about seventy Syrian SAM batteries plus an unknown quantity of AAA batteries waiting for them in full alert. They went into attack in a long stream which made it easy for the Syrians to adjust their fire. The Syrian radar operators used their radars only intermittently, which meant that not all the ARMs could be launched. During the fourteen or fifteen minutes that the attack lasted, at least forty missiles were launched and countless rounds of AAA ammunition fired against the US aircraft. When the US formation set course back to the carriers, one A-6E and one A-7E were found to be missing while another aircraft was flying with its jet pipe partially destroyed by an IR-guided missile. The pilot of the A-7E which had been shot down had managed to eject from his aircraft and had been picked up from the sea near Jounieh by a Lebanese navy patrol boat. The pilot of the two-seat A-6E was killed and the navigator was taken prisoner, but was released on 3 January 1984 after a month in prison.

According to US Navy officials, the mission achieved the result of preventing further Syrian attacks on US reconnaissance aircraft. Nevertheless, the loss of the two-aircraft was felt by many to be a humiliation for the US Navy, especially after Israeli aircraft carried out two similar attacks in the Lebanon, on 3 and 4 January 1984, with no casualties. Twelve aircraft took part in the first of these raids, on the Islamic Volunteers training camp at Baalbek; only four aircraft, Israeli-built IAI Kfir attack aircraft, actually carried out the attack, while the other eight provided air cover and ECM escort (jamming of enemy radars and the launching of chaff and IR flares). Sixteen Israeli aircraft took part in the second raid and this time only two carried out the actual attack while the other fourteen were used for long range escort and for surrounding the attackers with a protective barrier of chaff, heat balloons and IR flares.

It was inevitable that the American raid should be compared to those carried out by the French and the Israelis; there was almost unanimous agreement that the difference lay not in the expertise of the pilots, nor in the performance of their aircraft nor the tactics employed, but solely in the countermeasures used. According to US

Navy officials, the cause of the loss of the US aircraft and the damage inflicted on the third was that the Russians had modified the IR sensors of their ground-to-air SAM-7 and SAM-9 missiles. More precisely, in fact, it was the wavelength and filters of the sensors which had been changed, enabling the missiles to home onto the exhausts of the aircraft without being decoyed by the IR flares launched by the Americans. These modifications made by the Soviet electronics industry were, however, predictable given the high susceptibility of SAM-7 missiles to IR countermeasures shown in all conflicts from the Yom Kippur war onwards. The cause of this susceptibility to IR flares was that the uncooled heat seeker was easily 'pulled out' of its lock-on to the aircraft's heat signature, being attracted by the greater heat produced by the IR flare. The new SAM-7B version has the following characteristics, according to US Navy experts:

improved filters better to distinguish a real from a false target

a different wavelength

a more sensitive cooled heat-seeker.

These improvements were also made to SAM-9 *Gaskin* missiles, which are mounted on a vehicle in groups of four, unlike the SAM-7 which is a man-portable shoulder-launched missile. To enhance the acquisition capability of the new SAM-9 arms system, a *Gun Dish* radar was mounted on the vehicle endowing the system with remarkable precision. This well-known fire-control radar is used to control the Russian ZSU-23-4 *Shilka* system of four-barrelled 23 mm cannon.

Furthermore, it is probable that the Syrians deployed not only the whole range of Soviet SAM systems, from the SAM-2 to the SAM-9, but also the latest SAM-10, SAM-11 and SAM-13 systems designed to replace the SAM-4, SAM-6 and SAM-9, respectively.

To counter such an impressive display of AA arms, the US A-6E and A-7E did, nevertheless, have a range of ECM systems just as respectable as those of the aircraft of the Israeli air force. Presumably, the US A-6E and A-7E had the following equipment: two AN/ALE-39 Countermeasures Dispensers each capable of launching thirty RR-129 chaff units and thirty IR flares; an AN/ALR-45 RWR with automatic chaff-launching; and an ALQ-126 deception jammer, a later version of the ALQ-100.

However, what the Israeli aircraft did have and the US A-6Es and A-7Es did not have was a simple infrared countermeasure which consisted in lengthening the jet pipe of the engine so that a missile would detonate at a point sufficiently removed from critical structures to reduce resulting damage to survivable proportions. The Israelis also used this device on the General Dynamics F-16 Fighting Falcons they received from the United States. Unfortunately, it cannot be applied to carrier-borne aircraft since, with such long jet pipes, aircraft would not fit in the lifts used to transfer them between the hangars to the flight deck.

However, the Americans did have an IRCM device, the AN/ALQ-123, which deceives a missile's infrared seeker reticle. IR deception operates in a similar way to EW deception of a conical scan radar. In practice, an IRCM device emits a suitably modulated emission of IR energy which, when combined with the heat (IR energy) emitted by an aircraft's jet pipe, creates an angular deception signal which causes the missile's seeker to aim in the wrong direction.

So far, however, neither the AN/ALQ-123 system nor IR flares have been effective enough to ensure the survival of the pilots of aircraft equipped with such IRCMs, because the pilot never knows exactly when to activate the systems. Each aircraft can carry only a limited supply of flares which may, in certain circumstances, already have been exhausted before the point of greatest danger is reached. The other system, active deception, can have counter-productive results if activated too soon or at the wrong moment.

The only solution to these problems is an effective infrared warning receiver (IRWR), but it seems that the main defect of IRWRs—that of false alarms—has yet to be overcome by the Americans. Modern IRWRs are extremely effective in detecting a source of heat but, unfortunately, this heat can come from any nearby source, not only a missile, and the result is an unnacceptably high number of false alarms (the same problem affects IR homing missiles). Few nations have managed to solve this difficult technological problem and those which have naturally keep their findings top secret.

In the first analysis, one can safely say that the two American aircraft were shot down and the third damaged, once again, because of shortcomings in their protective EW systems which were not sufficiently advanced to counter the latest threat. Even the United States of America, which spends astronomical sums on defence and possesses the most advanced technology in the world, found themselves imperfectly prepared in the Lebanon as far as IRCMs were concerned, just as had been the case with ECMs in the Vietnam war.

New lessons from the Falklands and the Lebanon

The events so far described once again show that the face of warfare has changed, in that new elements and new operational concepts have come into play in the military arena.

The first new element derives from computer technology. The Royal Navy, with its centuries-old tradition of glorious naval warfare, owes its success in the Falklands also to the excellent organisation of command, control, communications and intelligence. The integration of these disciplines has been designated C^3I in the West, the exponential expression (C *cubed* I) being used to underline the multiplied effectiveness gained by joining together and strengthening the individual networks of command, control and communications.

In air, naval and ground warfare, command decisions must be fast and based on accurate information. Modern computerised C^3I systems are able to furnish, in 'real time', those responsible for making decisions with all the intelligence needed to provide an overall picture of the situation so that they can make a rapid evaluation and take the most appropriate action to deal with the threat or threats.

The Israelis have also exploited C^3I systems in their air and ground operations with excellent results. Particularly, in their coordinated use of aircraft and RPVs for reconnaissance in 'real time' and in their attacks on the Syrian SAM bases in the Bekaa valley. C^3I systems enabled the Israelis to acquire all relevant information about the enemy and pass it on without delay to those in command.

Another element which contributed greatly to the Israeli success in the Lebanon was the coordinated use of ECMs against enemy command, control and communications systems. These counter-measures, which are not only electronic, are designated C^3CM.

In all the conflicts between the Arabs and the Israelis in the last ten years, losses have been in favour of the Israelis, but never to such a degree as in their battles against the Syrians in the Lebanon. A ratio of aircraft losses of over 50:1, the suppression of the entire network of Syrian SAM systems in the Bekaa valley with very little damage to themselves, and the destruction of so many enemy tanks can be attributed to many factors, two being better tactics and better training.

However, this time there were extra elements. First, the wide-spread, coordinated application of ECMs and weapons systems against enemy command, control and communications systems. Just before or just at the outset of a battle, weapons systems and ECMs were used against enemy radars, communications networks and command and control centres with the result that the enemy was paralysed, unable to see, hear or communicate in any way. A clear demonstration was given of how to conduct Electronic Warfare, taking maximum advantage of the concept of C^3 Countermeasures (C^3CM) and fully exploiting 'real-time' reconnaissance. The Israelis demonstrated the highest degree of coordination between RPVs, guidance and control of their own aircraft using E-2Cs, jamming of radar and communications using the Boeing 707 and, finally, the actual means of material destruction themselves, such as aircraft armed with anti-radar missiles or RPVs, full of explosive, targetted against enemy radars and SAM-6 batteries.

If C^3CM systems continue to progress at the rate they have been doing in recent years, the point will soon be reached where battles can be won before they have even begun! It will be sufficient to use ECMs and appropriate arms (anti-radar missiles, anti-antenna munitions and so on) just before the battle is about to begin to paralyse the 'brain and central nervous system' of the enemy,—his C^3I organisation. In addition, the use of computers and micro-processors has also revolutionised fire-control and missile guidance technology, making modern weapons increasingly accurate.

Events in the Falklands and the Lebanon have also shown that Airborne Early Warning radar (AEW) will be indispensable in future air battles. The Israeli's intelligent use of E-2C Hawkeyes to detect and track the Syrian fighters as soon as they left their bases, together with the use of passive ESM systems to locate and identify enemy radar and missile installations are two more important factors in their success. In contrast, the British sorely lacked an Airborne Early Warning capability during the Falklands war and, consequently, their ships could be surprised by enemy aircraft and missiles.

The importance of RPVs for the purposes of Electronic Warfare has also emerged from the operations in the Lebanon and these will no doubt have more widespread use in future.

Today, a Task Force must have an integrated defensive coverage composed of systems covering a series of concentric circles of decreasing radii. Proceeding from the outer perimeter to the centre of the Task Force, these circles contain the following elements:

Airborne Early Warning radars (AWACS, E-2C, etc.) linked to combat air patrol or interceptor aircraft to shoot down enemy aircraft before they are within range to launch their missiles or expend their bombs surveillance systems, installed on ships and/or helicopters deployed some distance away from the main body of the Task Force and consisting of ESMs for detection and identification of the threat; naval radars to search for the threat and for target designation; jammers to be used against enemy search radar and to prevent, or at least delay, attacks by enemy aircraft

long and medium range SAMs and AAMs to be used against any aircraft that manage to penetrate the above defences

ECM helicopters to create artificial false radar targets and dispense decoy chaff and flares

Infra-Red Support Measures (IRSM) to detect low-flying missiles and/or aircraft via heat-emission, and for precise arms designation

anti-missile deception jammers, anti-missile missiles and expendable missile anti-homing deception jammers for use by ships under missile attack

modern anti-missile guns (Close In Weapons Systems—CIWS).

Ships of the *Sheffield*'s class, Type 42 destroyers, could certainly have been equipped with more advanced Electronic Warfare systems than they actually were, since such systems were already, in fact, in production by the British defence industry. The reason why such ships were not equipped with the latest equipment is perhaps that, because of the usual cuts in the British Defence budget, the Royal Navy thought they would instal their latest EW equipment only on new ships of later construction, and not retrofit existing classes.

The Task Force ships also lacked modern passive IR systems for the detection of aircraft and, above all, low-flying 'sea skimmer' missiles. A modern, panoramic IR search system would certainly have enabled the British ships to detect and locate Exocet missiles, even at extremely low altitudes, since IR systems, unlike radar, are immune to clutter.

Events in the Falklands teach the important lesson that, in order to detect the deadly threat of modern missiles, ships can no longer rely solely on radar and ESM on board but must have recourse to all the latest findings and improvements which modern technology places at their disposal. In addition to the aforementioned IRSMs, the fleet must be equipped with modern, expendable ECM systems—for example, deception jammers launched by missiles against an anti-ship missile—and helicopters dedicated to EW which, taking advantage of their mobility should place themselves between the threat and the ships threatened, jamming or deceiving the missiles and their command platform.

It emerges, from all the air and sea battles fought in the Falklands, that the anti-ship aircraft-missile combination in this case (the Super Etendard-Exocet) has truly revolutionised sea warfare. However, no definitive conclusions can be drawn about the performance of this combination versus warships off the Falklands since, in the final analysis, most of the British ships lost in the South Atlantic sank due to inadequate damage control and fire-fighting systems.

The Falklands war, like those modern wars examined previously in this book, also brought to light shortcomings in the field of intelligence. Despite clear signs of imminent Argentinian action against the Falklands, Great Britain failed to foresee the invasion. It would have been sufficient to send nuclear-powered submarines and a number of aircraft and ships to dissuade the Argentinians from an undertaking which, as the events themselves have shown, was destined to fail.

The gravest error on the Argentinian side was that of underestimating the British reaction. They deluded themselves in believing that the *fait accompli* would sooner or later be accepted by the British who, in the last ten years, had divested themselves of more important dependencies than the Falklands, and who were on the verge of making further drastic reductions to their fleet.

However, at the tactical level, great care was taken with intelligence, especially by the British, and it was one of the factors which most influenced the outcome of the battles in the Falklands. In fact, drawing on a variety of technical and other sources, the British Forces always had substantially better information regarding Argentinian force levels, deployment, tactics and intentions compared to the quality and quantity of information the Argentinians had regarding UK forces. One can accurately say that British intelligence had a strong influence on the outcome of ground battles in the Falklands and that the British campaign would have gone on much longer without its contribution.

In general, the experience of the Falklands demonstrated the necessity of having precise, up-to-date information regarding the performance of weapons and sensor (i.e. radar, IR, laser, etc.) systems of all world powers, whether potential enemies or allies. Above all, it showed the need to make a greater effort in collecting and analysing information regarding all potential threats and not only what is considered by the Western nations to be the greatest—the USSR.

Events in the Falklands and the Lebanon have also provided confirmation that ground warfare has changed substantially. In the Falklands, the British nearly always attacked by night, using night-vision devices and weapons especially designed for night fighting and employing tactics involving making quick, surprise attacks on the enemy without being seen. In the Lebanon, the Israelis employed modern anti-tank techniques which they had devised themselves and led to the destruction of a number of T-72 tanks—the pride of Soviet military industry.

Consequently, in order to survive, modern armies must make sure that their tactics are in step with the latest developments in electronic and electro-optical technology which, along with rigorous training and aggressiveness, are the winning factors on the battlefield.

In the air, the AIM-9L Sidewinder air-to-air missile was used with great success by the British Harrier fighters to shoot down Argentinian aircraft and by the Israeli fighters against Syrian aircraft. It represented a triumph for infrared technology. The ability of the AIM-9L version to hit an enemy aircraft from any direction (All-Aspect Capability—ALASCA), and not only from behind, as was the case with early AIM-9 versions, gave the British and Israeli pilots an enormous advantage which no amount of manoeuvring by some enemy ace could match.

However, as was shown by the shooting down of the US A-6E and A-7E by the Syrians over the Lebanon, the growing use of IR technology has also become the most insidious threat for the pilots of attack or close air support aircraft. The pilot of an aircraft on a penetration mission must be equipped with an IRWR (Infra-Red Warning Receiver) able to detect a missile not at the last moment but at the very moment of launch in order to give the pilot enough time to actuate appropriate countermeasures or commence an immediate evasive manoeuvre.

It is more difficult for a ship to detect an anti-ship missile because these are often launched from much greater distances than air-to-air missiles. Consequently, a ship must have an IRWR able to detect an approaching anti-ship missile flying at an extremely low altitude and which, due to its limited radar surface and the fact that its radar and navigation system may not emit, may completely escape detection by radar and ESM.

It seems likely that 'stealth' techniques will be used in the future to make aircraft and missiles invisible, or almost invisible, to radar. When this happens, search radar will necessarily have to be integrated with other systems such as IRWRs and other sensors using new techniques.

'Stealth' techniques are based upon several elements. In regard to the physical aircraft itself, it is essential to reduce its radar and IR signatures. Its radar reflectivity, including its scattering and diffraction properties, is a measure of its efficiency in intercepting and returning a radar signal—its radar signature—and depends upon the aircraft's shape, size, aspect and the dielectric properties at its surfaces. It can be reduced by using radar-absorbent materials (RAM) and appropriate geometries (shapes and configurations). As an instance of the dramatic advances in this field, using these techniques the Rockwell B-1B, the US strategic 'stealth' bomber and ASM platform due to enter service in 1987, has a radar cross-section one hundredth that of the comparably-sized Boeing B-52 and one tenth that of the B-1A prototypes of the 1960–70s! An aircraft's or missile's IR signature—that is, the heat emitted by the engine and jet-pipe—can be reduced by greater thermal efficiency and insulation, but is impossible to eliminate. In the field of equipment, use can be made of bi-static radar equipment, a special radar whose receiver is installed in the 'stealth' aircraft, but whose transmitter is located elsewhere, either on another aircraft or on the ground. Terrain-following radar, permitting very low-level, contour-hugging flight below radar cover to and from a target area, taking advantage of ground clutter, is standard 'stealth' equipment, in use on, for instance, the RAF's Phantom FGR.2 (F-4M) tactical aircraft in the early 1970s, and its successors, the Anglo–French SEPECAT Jaguar GR.1, purpose-designed for low-level operations. In operational terms, a great deal can be done to protect aircraft by 'stealth' techniques, notably using terrain-following radars and selecting the most suitable mission profiles to propagation conditions i.e. those pertaining to sending out radiation such as infrared and radar.

Again, regarding air operations, it must be observed that air losses in the Falklands, especially on the Argentinian side, and in the Lebanon, particularly by the Syrians, were far too high in proportion to the number of aircraft employed. This can be explained mainly by the fact that nearly all Argentinian and Syrian aircraft—and quite a few British aircraft—lacked active ECMs—jammers and electronic and IR deceivers—which made it extremely difficult for them to escape the numerous types of anti-aircraft missiles used in these conflicts, such as Sea Dart, Sea Wolf, Sea Cat, Rapier, Roland and Sidewinder. All these considerations confirm, as we have seen in previous conflicts, that electronic and IR systems are essential factors in reducing air losses.

In conclusion, the conflicts in the Falklands and the Lebanon represent an important turning-point in the history of war because they demonstrate that Electronic Warfare is an irreplaceable instrument of success both in offensive and defensive operations. In particular, the battles fought in the Lebanon have proved beyond any shadow of a doubt that the result of future battles will depend much less on the quantity of the aircraft, ships or tanks used than on their quality, which naturally includes new developments in the field of electronic technology.

However, Electronic Warfare is not static: the mere possession of a certain number of ESM or ECM devices is not enough to ensure success. In Electronic Warfare, what works today may not work tomorrow, and developments in EW systems must always closely and appropriately follow developments in the threat. With the endless evolution of applied military technology, electronically-guided weapons are coming closer and closer to perfection and thus constant up-dating and refinement of EW equipment is required.

The extremely dynamic and evolutionary character of EW also, unfortunately, demands constant, heavy financial expenditure. If a potential enemy changes the frequency of one of his radars, develops a new anti-jamming device or makes some important change in the IR guidance system of a missile then the potential opponent has to modify or even completely renew his own EW equipment. However, from any point of view, this is certainly the most worthwhile investment for armed forces.

When the red light of an RWR or IRWR illuminates in the cockpit of a warplane or the CIC of a warship on a mission, it means that, within seconds, a missile will hit that aircraft or ship if nothing is done. In those instants, the lives of the entire crew, the survival of the aircraft or ship and the success of the mission itself depend almost entirely on immediate identification of the missile and actuation of appropriate countermeasures. Only if the EW and IR devices can completely and immediately deal with the imminent threat, will the crew's survival and the mission's success be largely ensured.

If providing aircraft and ships with the electronic means to greatly improve their chances of survival and success appears to be an arduous and costly task, one must bear in mind that the cost of a modern warplane, warship, tank or, even more so, trained crew is very much greater than the cost of providing them with EW equipment. This investment must be made in peacetime because the price to be paid once an unexpected war has broken out will be extremely heavy.

It is, therefore, vital that intelligence is gathered continuously in peacetime by SIGINT and IRINT operations and by other means in order to acquire information regarding the parameters of new weapons systems deployed or under development by potential enemies. Finally, emphasis must be placed upon providing national facilities for technical and scientific research in order to develop the technology necessary to achieve and maintain superiority in Electronic Warfare which has now become an obligatory route to success and the survival of a country's armed forces.

23

Electronic Warfare in Space

Man's conquest of space has brought a new dimension to arms technology, communications systems and methods of surveillance with consequent innovations in the field of EW.

The use of satellites for military purposes began in 1958 when the United States launched the communications satellite Score which simply transmitted pre-recorded messages from space. Ever increasing numbers of satellites, initially experimental and later operational, were launched in the following years with the intention of establishing a comprehensive communications system in space. To solve the complex problems of command, control and communications which might arise in the event of a world war, the United States set up a truly world-wide satellite communications network providing secure and effective radio communications, between central commands and military units deployed in any, even the most remote, part of the world. The system would have to be immune to atmospheric disturbance, local interference and, above all, jamming and deception (ECMs).

At a distance of 36,000 kms, each satellite in this communications net is in geostationary orbit, that is, it is apparently motionless above a fixed point on the Earth's surface. Each covers an area equal to a third of the Earth's surface and special techniques are employed to reduce the susceptibility of the signals transmitted to all forms of interference and jamming. Today, this system ensures immediate radio contact between two military commands separated by thousands of miles with the same clarity and efficiency as radio contact between two ships a few miles apart. It has also solved the problem of transmission of orders to submarines armed with SLBM (Submarine Launched Ballistic Missiles) with nuclear warheads, such as 'Trident' and 'Poseidon'; these form a vital part of the US nuclear reprisal 'triad', the other two 'prongs' being ICBM (Inter Continental Ballistic Missiles) and manned bombers. The problem, for which traditional communications had been unable to offer a satisfactory solution, was to ensure that these submarines would receive their nuclear orders with absolute certainty but without themselves having to transmit radio signals which carried the inherent risk of interception.

Since launching ballistic missiles from submarines requires

285

extremely accurate knowledge of the precise launch position, which no navigation system then in existence was able to provide, the Americans decided to use satellites to provide precision navigational information. The first satellite of this type, called NNSS-Transit (Navy Navigation Satellite System) was sent into orbit in 1960 and fulfilled most of the requirements of US nuclear submarines.

The United States has recently been deploying a navigation system called Navstar GPS (Global Positioning System) consisting of twenty-four satellites in orbit at a distance of about 20,000 km from Earth which continuously emit special signals enabling ships, aircraft, or even infantry equipped with special receivers to fix their position with amazing accuracy—a margin of error of about ten metres!

It is, however, in the fields of photographic and electronic reconnaissance and Early Warning that artificial satellites have proven most useful for military purposes. The first reconnaissance satellite, *Discoverer*, was sent into orbit by the Americans in February 1959. It represented a completely new way of conducting reconnaissance since aerial photographs of enemy territory were taken without direct human intervention and from a distance far beyond the range of any weapon on Earth.

The serious episode of the shooting down of Captain Francis Powers' U-2 spy-plane over Russia in 1960 was largely responsible for convincing the US administration of the need to speed up development and production of this type of satellite in order to be able to proceed in safety with the task of gathering precious and indispensable information for EW which had previously been accomplished by means of missions such as that being carried out by Powers.

Since then, thousands of increasingly sophisticated and varied reconnaissance satellites, commonly known as spy-satellites, have been launched, mainly by the Americans and the Russians. In May 1972, in the course of SALT 1 negotiations, a special agreement was reached between the Russians and the Americans—the so-called 'open space' agreement— so that, today, reconnaissance satellites are one of the few internationally accepted means of acquiring information. Such use is accepted for 'verification' of the SALT and START treaty provisions.

Spy-satellites have similar functions to those of U-2 spy-planes. Both carry high resolution photographic cameras as well as equipment for intercepting and recording electromagnetic emissions present in the skies of potentially hostile countries i.e. all communications and radar signals emitted by the electronic devices of these countries. The difference lies in the fact that, while the U-2 brings the films and tape-recordings back to Earth, the satellites transmit the pertinent data, duly coded, directly and instantaneously back to the receiving stations for immediate analysis, facilitating 'real-time' appreciation of the information.

Reconnaissance satellites are placed into various types of orbit, the duration of which depends on the type of reconnaissance to be carried out. When their particular mission comes to an end, they disintegrate and burn-up on re-entry to the Earth's atmosphere.[1]

For improved coverage, two satellites are often launched simultaneously into parallel orbits but at different heights, the satellite in the lower orbit photographing an enemy radar installation which the one in the higher orbit had discovered by interception of its electromagnetic emissions. Spy-satellites have also been launched which, after having picked up the signal of a hitherto unknown radar, are able to reposition themselves lower during later orbits to the optimum altitude for taking photographs.

In recent years, the series of US reconnaissance satellites has shown continuing technological and operation progress. One of the first reconnaissance satellites used by the United States was the SAMOS (Satellite and Missile Observation System). Until a few years ago, SAMOS satellites were launched in great secrecy by an Atlas rocket at monthly intervals into orbits which criss-crossed the skies over the Soviet Union. At set times, special capsules were expelled by the satellite containing exposed film and the tapes of EMG signals which had been recorded. Slowed by deployment of little parachutes, these fell in a preselected zone in the Pacific Ocean where they were picked up by the numerous US warships standing by to pick them up. However, there were always large numbers of Soviet 'fishing trawlers' (i.e. spy ships) also waiting in the same zone where the capsules were to be dropped and so the Russians sometimes managed to get their hands on the material before the Americans! For this reason, aircraft were adapted for aerial pick-up of the cassettes for more secure recovery of the material.

Besides the SAMOS satellites, the Americans also used MIDAS satellites which were fitted with numerous IR sensors for IR surveillance of the Soviet Union. Every time the Soviets launched an experimental ballistic missile, the IR energy emitted by its engines was automatically picked up by the MIDAS satellites.

Later, an even-more sophisticated satellite was developed, called Big Bird, which weighs over 11 tons. Big Bird's incredible operational performance has enabled the United States to keep track of the latest technical and operational developments of a military nature in the USSR and other potential enemies. The Big Bird is able to take exceptionally clear photographs, develop them and transmit them to Earth in code. It can also carry small electronic reconnaissance satellites which are expelled and placed in independent orbits to record the emissions of any new radars which are discovered.

On 11 May 1982, a Big Bird was launched from Vandenberg Air Force Base in the USA into an elliptical orbit of 169–257 km and a polar inclination of 94.4 degrees. This provided vital information which was passed to the British during the Falklands conflict. Observation of Argentine troop movements, tracking of Argentine naval units and surveying via high resolution camera of possible sites for a British landing (including the actual site at San Carlos) were all undertaken, and the results passed to the UK where the data was transmitted in 'real time' to the British Task Force via satellite link to their shipborne Scot Skynet terminals.

The Americans have stationed an ELINT reconnaissance satellite of type Ryolite in geostationary orbit at an altitude of about 36,000 km and directly above the Soviet missile test range which runs from the launching complex at Tyuratam to the landing area in the region of the Kamchatka Peninsula (see opposite). Its IR sensors are alerted to a launch by the heat emitted. They also operate 'listening posts', installed under agreement with the People's Republic of China, at Korla and Qitai in the mountainous region of Xinkiang in Northwest China, as well as their stations at Shemya and Adak in the Aleutian Islands. All of these are intended to gather telemetry signals on Soviet missile tests; such signals contain data on missile performance, warhead or multiple warheads, and Circular Error Probable (CEP) i.e. accuracy.

In another category in the US inventory are the Early Warning Satellites whose main task is to prevent the possibility of surprise attacks. To this end, they carry a variety of ultra-sensitive sensors capable of discovering nuclear explosions in any part of the world, detecting the launching of ICBMs, undertaking passive IR sur-

The Electronic Warfare environment in North-East Asia. Note the trajectory of Soviet missile tests from Tyuratam to the region of the Kamchatka Peninsula and the US 'listening posts' for interception of their telemetric emissions. It was into this 'dense' electronic environment that the Korean Airlines Boeing 747 had strayed when it was shot down by the Soviets in 1983.

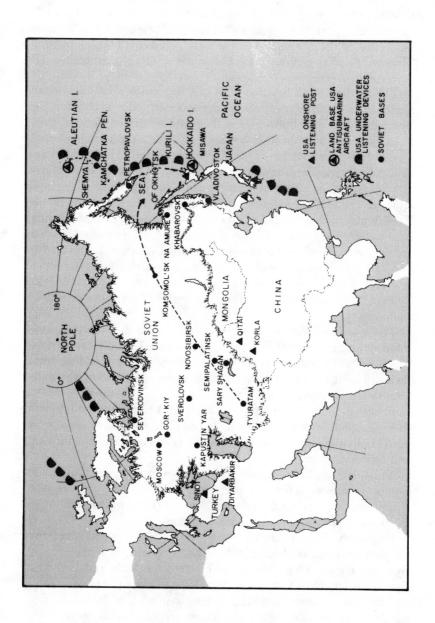

veillance of any source of IR energy, including explosions, fires and new, sometimes disguised industrial plants, and carrying out oceanic surveillance. The latter activity consists in keeping track, even in peacetime, of the movements of hostile or potentially hostile ships and submarines all over the world. Ocean surveillance satellites use IR detectors and other types of sensors.

The Soviet Union has also been carrying out similar missions in space, their satellites undertaking the same activities as those of the American's. Nearly all those for military use are named Cosmos, their exact functions and purposes not being specified. Over one thousand Cosmos satellites have been launched by the Russians so far. They are of widely differing sizes and cover a wide range of functions: some have also been used for scientific research such as exploration of the high strata of the atmosphere and outer space and measurements of the Earth's magnetic field, and solar radiations. However, most Cosmos satellites are for specific military purposes. Those for reconnaissance are sent into orbit from the missile ranges of Tyuratam, Kapustin Yar and Pelsetsk and usually carry small capsules which are ejected after a few days. Some of these satellites are fitted with a motor which enables them to manoeuvre over objectives to be explored and move in or away as necessary.

The Russians, like the Americans, make sure that precious little information regarding their military satellites leaks out. Nevertheless, it is possible to deduce the target zone of their reconnaissance from their orbital inclination (i.e. the angle between the plane on which the satellite's orbit lies and the plane of the equator). It is also possible, by statistical analysis of the number of satellites launched, their launch dates and of other features which cannot be kept secret, to make valid predictions about their tasks and equally valid inferences of a political and military nature.

Nearly all satellites launched by the Soviets fly over the United States. It can therefore be presumed that, besides photographic equipment, they are also equipped with receivers which pick up all electromagnetic signals held to be of interest.

During the Indo–Pakistani war in 1971, the Soviets launched two Cosmos reconnaissance satellites to follow operations in that part of Asia. During the Yom Kippur war between the Arabs and Israelis in 1973, they launched two Cosmos satellites to carry out photographic and electronic reconnaissance for the Egyptians and later launched another five to check that the cease-fire conditions were being observed and Israeli forces were being progressively withdrawn.

At least a hundred of the Cosmos satellites launched so far have been for electronic reconnaissance in general and to gather electronic information on NATO radars in particular. Like the Americans, the Russians use some of their Cosmos satellites to maintain a constant watch on the various US fleets (the seventh in the Pacific, the sixth in the Mediterranean, the second in the Atlantic, etc.) and to monitor their positions in order to be able to guide their destructive weapons against them in the event of war. These satellites often have nuclear-powered radar as proved to be the case when fragments of Cosmos 954 fell in Canada, causing great alarm at the prospect of nuclear contamination.

Both during and after the Russian intervention in Afghanistan and the US raid in Iran to free the hostages, the Soviets launched numerous reconnaissance satellites, the last ones being Cosmos 1179 and 1180, sent into orbit in May 1980. Cosmos 1180, unlike its predecessors which were obviously for reconnaissance over the Middle East, aroused considerable curiosity and suspicion as its parameters (e.g. inclination of orbit) were unlike anything seen before.

Soviet activity in this field is extremely varied and has included the most amazing missions of electronic espionage. For example, a Cosmos satellite was kept in orbit over Iran for a few years before the Islamic revolution, receiving information transmitted by undercover agents in Iran regarding troop movements and upon defensive installations along the border with Russia. This activity was brought to a halt when the Iranian counter-espionage service arrested a Soviet agent who was caught red-handed transmitting such information via radio directly to the Russian satellite.

According to US sources, there was a similar case in San Francisco, California in a zone where the EW industry flourishes. Since the companies operating there are forever coming up with new so-phisticated equipment, the Russians decided that it would be a good idea to send a Cosmos satellite into orbit over California to get information regarding such developments directly from source. This new form of electronic-industrial espionage was carried out in such a way as to not arouse suspicions.

On the roof of their Consulate building in San Francisco, the Soviets installed extremely sophisticated electronic devices capable of picking up even the weakest radio signals. In the southern part of the city, several miles away from the Consulate building, is the so-called 'Microwave Tower' where a radio repeater system is installed. This is used by the various local electronics companies to exchange communications regarding research, development, production and testing of new special microwave components, such as semi-conductors, and TWTs (Travelling Wave Tubes), which, in their respective technological fields, are the most advanced in the world. All these communications, with their precious information on the construction of EW components and equipment, were recorded by technicians at the Soviet Consulate. The data was prepared and then passed on to a KGB agent who, using a special portable transmitter, transmitted this information, always from locations a good distance away from San Francisco, directly to the satellite, at pre-arranged times. The electronics companies finally discovered the 'leak' and applied a series of ECMs to prevent further Soviet interception.

There have been many cases of industrial espionage, a common activity nowadays thanks to satellites and their EW equipment. The 'computer war' is another example, although little is said about it as few people realise how vulnerable computers are to ECMs. Many large industries use computer centres to solve design problems; satellites are often used as data-links between companies and distant computer centres, often internationally. By intercepting these communications which are frequently inadequately coded, it is possible to gain access to an incredible quantity of data regarding plans, which are often secret, for the research and development of defence systems in various countries.

Of course, every type of US satellite has its Soviet counterpart. There are also Soviet satellites whose functions are totally unknown in spite of being kept under constant observation by the Americans. However, the Russian satellites are, on the whole, less technologically and operationally advanced than the American satellites and have so far had a shorter 'operational life'.

Soviet military leaders decided to circumvent the obstacle of US satellite superiority, which would give the Americans a decided strategic advantage in the event of war, by embarking on an intensive programme to develop anti-satellite weapons. Bearing in mind the natural vulnerability of space vehicles, the Soviets designed and developed a new type of satellite capable of attacking and destroying any hostile or potentially hostile vehicle in space. Special techniques were also devised for rapid interception, neutralisation, or destruction of enemy military satellites, especially the USA's 'Early Warning' satellites which are responsible for detecting ICBM launchings.

The first Soviet experiments in this field took place in 1968. On 10 October, Cosmos 248 was sent into orbit, followed ten days later by

Cosmos 249, launched from the Soviet missile range at Tyuratam. Under full ground control, Cosmos 249 was launched into an orbit to intercept Cosmos 248 and then manoeuvred close up to its target, at which moment it was exploded, gravely damaging Cosmos 248. Thus the first 'satellite-killer' or 'anti-satellite satellite' was born.

Since then, the Soviets have made at least another fifteen test interceptions in space, all closely monitored by US space command and control stations. As soon as the Americans realised that the Soviets were in the process of devising a satellite-killer system capable of eliminating their surveillance and communications satellites and thus putting at risk the US nuclear deterrent, they immediately set about trying to remedy the situation. First, they built satellites with a special protective armour to defend them from the fragments produced by the explosion of Soviet satellite-killers. They also sent satellites into much higher orbits which the Soviet satellite-killers were unable to reach.

For a few years the Soviets made no further interceptions, perhaps to evaluate the results of tests which had been carried out; when they recommenced their test programme, a new technique was used. The interceptor-satellite, after making a few orbits, was brought in close to the 'victim-satellite', which it would track, travelling at the same speed but in a slightly lower orbit. Over a long period of time, the victim satellite was kept under observation in order to find out and transmit to Earth details of its function and main characteristics. Then, the interceptor would be guided towards the lower strata of the atmosphere where it would disintegrate. This new method, which the Russians began to adopt in 1976 and which obviously did not have the aim of destroying the victim satellite, was probably connected with experiments on new laser weapons.

At first, Soviet satellite-killers only intercepted other Soviet satellites, such as Cosmos 803 which was intercepted by Cosmos 804 on 16 February 1976 and by Cosmos 814 on 13 April of the same year. It is almost certain that another satellite-killer, Cosmos 843, sent into orbit in 1976, failed to intercept its target, Cosmos 839. On 27 December 1976, Cosmos 886 was clearly observed approaching its target, Cosmos 880, up to a distance of less than 2 km.

However, the following year there was growing concern in the United States over the fact that the IR sensors of two USAF satellites, used for retransmission of data required for wartime operations by Strategic Air Command's bomber force, were often temporarily blinded, especially when flying over Russia. On two occasions, 18 October and 17 November 1977, these two satellites, as well as other

US Early Warning satellites, were put out of action for almost four hours. CIA experts suspected that this black-out was due to deliberate jamming by the Soviets using a laser, either based on the ground or on a killer-satellite which they were testing.

The Russians continued their anti-satellite tests in 1977, using increasingly sophisticated techniques. The target-satellite Cosmos 909, established in an orbit nearly 2000 km high, was intercepted first by the satellite-killer Cosmos 910 on 23 May and then by the Cosmos 918 on 17 June. The latter, launched on 17 June 1977 from the Tyuratam range, was first placed in a very low orbit but subsequently manoeuvred into a much higher orbit to intercept the target-satellite Cosmos 909 at the same altitude as many US navigation and reconnaissance satellites. After interception, both Soviet satellites, the interceptor and the target, flew close together for a while and then headed downwards to disintegrate in the denser strata of the atmosphere.

Also in 1977, sixteen Soviet Cosmos satellites were launched as part of a space programme officially intended to measure the dimensions of the Earth and to make a more accurate study of the movements of the Poles and the so-called continental-drift phenomenon. However, according to US experts who observed the orbits of these satellites, their real functions were quite different. It would seem that the aim of most of these missions was to acquire precise data regarding the position of important targets in the United States and Europe so that, in the event of war, similar Cosmos satellites could be used to guide Soviet ICBMs to these targets.

Other missions were set up to test new techniques for intercepting, approaching and destroying the enemy satellite and then effecting re-entry. The technique of detonating the satellite-killer itself to destroy the enemy satellite was abandoned by the Soviets, perhaps because it was necessary to manoeuvre the interceptor-satellite into very close proximity to its target and in very few of the experiments did they manage to get closer than 1 km. Given the total lack of atmosphere, an explosion in space causes no shock-wave effect which is what causes the most damage in explosions within the atmosphere. Consequently, the Soviets changed the direction of their research to the field of 'soft-kill' methods of satellite neutralisation. ECMs were tried out to ground command transmissions to satellites essential for keeping them in the desired orbit or to deceive the satellites by giving them false commands to descend into the low atmosphere where they would burn up. However, it would seem that these systems have now also been abandoned because the Americans have equipped their satellites with ECCMs, including coding of commands and incorporating anti-jamming and anti-deception devices.

In the belief that, for certain purposes, neutralisation is just as effective as destruction, the Soviets have developed new methods to be used after the satellite-killer has approached its victim, possibly in that part of the orbit which cannot be observed from enemy Earth stations. One such method involves the use of manned-spacecraft, whose crew actually board the enemy satellite. The astronauts leave their spacecraft and render vital parts of the satellite inoperative, either directly or by remote control, using various types of radiation or corrosive chemical substances, or even removing vital parts without which the satellite's functional efficiency would be greatly degraded. Finally, small rockets can be attached to the satellite either to accelerate it and thus send it into a more distant orbit or to slow it down and thus allow the Earth's gravity to drag it into the lower strata of the atmosphere where it will burn up.

Fearing that the Soviets, in the event of an impending conflict, would neutralise enemy satellites in orbit, the Americans, initially, developed a series of jammers to counter an electronic attack. However, the task of jamming satellites in orbit is fraught with technical problems which cannot easily be solved and so they finally opted for passive ECMs such as chaff and false IR targets capable of deceiving a killer-satellite.

The sphere of action of ECMs in space is not limited to anti-satellite operations, however; it has also begun to embrace the study of possible actions against ballistic missiles. Generally speaking, it can be assumed that many EW actions carried out in the atmosphere against radar and IR guidance systems would also function in space against the sensors of a ballistic missile, or, more precisely, against its nuclear warheads which, in the last phase of the trajectory, separate from the missile-carrier. Deception has recently become a factor of growing importance in ICBM tactics for both offensive and defensive purposes. Experiments have been carried out using IR-guided missiles against ICBMs in the phase when the ICBM is still under rocket propulsion. The anti-missile, by homing onto the heat produced by the combustion of the ICBM's 'booster', may cause it to explode prematurely. In the United States, anti-ICBM experiments have already been carried out in the atmosphere.

IR decoys have proved to be particularly effective for protection against ICBMs. By creating powerful sources of IR energy of the same wavelength as that sought by the ICBM or its nuclear warheads, they are able to divert the deadly weapon from its true target.

However, the Americans still mainly rely on their Early Warning satellites to instantaneously alert them to Soviet ICBM launchings against the United States in order to avoid surprise attacks. Therefore, to protect these satellites, the Americans developed radar and IR warning receivers for installation on the satellites to provide them with early warning of the approach of a hostile satellite, allowing them time to manoeuvre away dfrom it. This anti-satellite-killer or satellite-killer-killer system possesses the means of surviving an attack and informing ground control in the event of neutralisation so that a reserve satellite can immediately be sent into orbit to replace it. The Americans keep a number of reserve satellites ready for immediate launch to ensure that vital services provided by satellites are not broken off by enemy action.

Having fallen far behind the USSR in anti-satellite development, the USA produced a series of ECMs to degrade Soviet reconnaissance satellite performance—particularly those used for tracking US Naval surface ships and submarines. They also embarked on a crash programme, one development of which is the ASAT (Anti-Satellite) missile which is launched from a high-flying McDonnell-Douglas F-15 aircraft. They are also developing methods of using Space Shuttle manned craft to mount anti-satellite operations.

Meanwhile, the US intelligence services continued to supply information on Soviet progress in space warfare, a matter of growing concern both to the general public and government leaders. According to the CIA, the Russians had made a series of new tests using a high-energy hydrogen-fluoride laser developed at their research centre at Sary Shagan near the Chinese border and were in the process of transporting new extremely powerful generators there. The tests indicated that the Soviets were considering the use of powerful lasers against US satellites and ballistic missiles in space. To avoid the degrading effect that the atmosphere has on the laser-beam, small lightweight laser-weapons were under development for use on board spacecraft against enemy satellites.

Artist's impression of a future battle between high-energy lasers and ISBMs. In example one, an enemy ICBM, 1, reflected by a giant mirror, 2, orbiting in outer space, is located by a ground early warning station, 3. Immediately, a land-based high-energy laser emits a beam, which hits the incoming missile through a reflected ray from a low-orbiting mirror, 4, and the ICBM explodes, 5. In example two, an enemy ICBM, 6, is located by a satellite, 7, which attacks it directly and destroys it, 8.

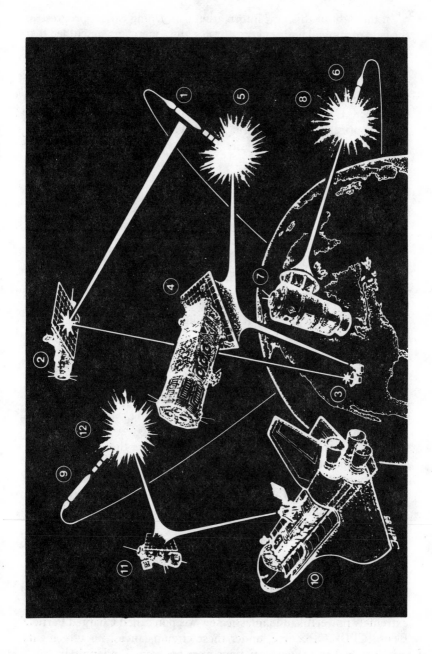

In the light of this new information concerning Soviet progress in the field of applied high-energy physics, the Americans concluded that the temporary black-outs of their satellites in October and November 1977 had been caused by Soviet experiments with high-energy laser weapons in space. This confirmed their fears that the Russians had, in fact, acquired the ability to temporarily neutralise their satellites. Moreover, US industry had not yet managed to develop any valid electro-optical countermeasure (EOCM) capable of intercepting and neutralising even a normal laser beam, mainly because of the high directionality of the beam itself.

As if this was not enough, US satellites then confirmed that tests had actually been carried out using a compact hydrogen-fluoride laser, capable of neutralising an enemy satellite at a distance of 1 km, at the large research centre of Krasnaja Pahka, 50 km south of Moscow. Moreover, preparations were being made to launch a similar laser-weapon on board a spacecraft. According to US officials, one such experiment had already been carried out from the manned Soyuz spacecraft.

Faced with such evidence, the Americans realized that they were at least ten years behind the Russians in the field of killer-satellites. Pentagon leaders came to the conclusion that the United States could not allow the Soviets to acquire and maintain a dominant position in space or to use their superior anti-satellite capabilities during an international crisis or in a direct confrontation to prevent the United States from using that vital element in their military system consisting of satellites for surveillance, early warning, navigation and communications.

In an attempt to gain time needed to complete their own research and development programme in this field, the Americans proposed negotiations with the Russians to suspend tests on anti-satellite weapons. Thus, on 8 June 1978, representatives of the two Superpowers sat round a table in Helsinki to discuss the problems of satellite warfare in space.

However, whereas the Americans came to the negotiating table full of proposals and good intentions, the Russians, on the other hand, arrived strong in the knowledge that their experiments with satellite-killers were proceeding successfully and that they were in the advanced stages of researching—in addition to high energy lasers—an extremely powerful and fantastic new weapon called Charged Particle Beam (CPB). Of course, under these circumstances, no agreements were reached and the negotiations were postponed indefinitely.

Soviet experimentation on the various systems for destroying or neutralising enemy satellites went on uninterrupted throughout 1979 and 1980 with great success. However, the news of this success was not announced by the Russians. According to US intelligence sources, in mid-March 1981, a Russian Cosmos satellite-killer had managed to completely neutralise the photographic, IR and electronic equipment of a US target-satellite, probably by means of a high-energy laser. The US report also suggested that the satellite-killer had employed special IR sensors to home onto its victim.

The Russians have never issued any information about experiments on their new CPB weapon either but numerous significant events related to such experiments were revealed by US spy-satellites orbiting over the areas where the experiments were carried out. It would seem that the new weapon, based on the principles of charged particle physics, is even more powerful than the high-energy laser. The difference between the two lies in the fact that, whereas laser uses photons, which do not have mass, the CPB weapon emits a stream of relatively heavy sub-atomic charges, such as electrons, with a negative electrical charge, and protons, with a positive electrical charge, at a speed close to that of light. These jets of energy do not melt the target as a laser does but smash it to smithereens. In other words, the CPB weapon works by thrusting the basic atomic particles of which matter consists—protons and electrons—at extremely high speeds against the target which is destroyed by the kinetic energy concentration and connected thermal effects. In effect, it is a kind of electronic gun, of unprecedented power and rather unusual shape, which emits pulses of jets of energy of the order of hundreds of billions of electron-volts.[2]

A CPB weapon is built by means of an 'accelerator' plant, a power generator, particle injectors and extremely high capacity condensors capable of storing very high power levels. Such machinery is generally very large and exceedingly complex.

On the ground, such a weapon would have a range of only 5 to 10 km due to atmospheric absorption but, if it could be installed on an artificial satellite and operated in space, its range would be increased to many hundreds of kilometres. The Russians probably first tested electron beam-emitting devices in space during the missions of Cosmos 728, launched in April 1975, and Cosmos 780, launched in November 1975.

News of Soviet development of these weapons first reached the rest of the world in 1975 when a US surveillance satellite detected the presence of large quantities of gaseous hydrogen containing traces of tritium in the higher strata of the atmosphere over Semipalatinsk in central Asia. This element is one of the necessary ingredients for creating charged particle beams.

US satellites also revealed that the Russians had transported a new, more powerful magneto-hydrodynamic generator to their research centre at Azgir in Kazakistan near the Caspian Sea for testing. This pulse-fuction generator would be able to supply the necessary power to operate a CPB weapon. A US electronic surveillance satellite orbiting over the Indian Ocean further revealed that a prototype of the weapon had been tested in a desert in the Azgir area.

The Americans have so far detected at least eight experiments on the propagation of particle beams in the ionosphere and outer space from manned space vehicles such as the Soyuz and unmanned Cosmos satellites. Observations have also been made of a series of experiments connected with the propagation of CPB against targets at the missile range of Sarova, near Gorki, to determine the effects of such propagation. Experts consider such tests to be the prelude to the development of a CPB weapon for use against ballistic missiles. Sarova, with its ultra-modern equipment for accelerating electrons, is held to be the headquarters of this research and development programme, which is directed by Professor M. S. Rabonovich of the Lebedev Institute in Moscow. A new accelerator has recently been built at Sarova and its extremely high power may eventually be used for the development of a proton beam weapon. The construction and testing of this accelerator were carried out under the direction of the physicist A. J. Pavlovskij.

Following these revelations, another discovery has been made regarding Soviet activity in the field of radiation arms. In early 1978, high levels of thermal radiation and the presence of nuclear waste discharges were noticed coming from the experimental plants at Semipalatinsk. These factors would seem to indicate that experiments to perfect the power sources for CPB weapons were being carried out. It has also been confirmed that an enormously powerful pulse generator has been built in a top secret zone near Sary Shagan to be used as a source of energy for such weapons.

The technical difficulties connected with the development and installation of a weapon of this kind on a space vehicle are so great that considerable perplexity has been expressed by US authorities about the usefulness of spending so much money on a weapon which would seem to be so difficult, if not impossible, to develop. Nevertheless, the report by the Chinese authorities, mentioned previously, according to which many Chinese soldiers had been hospitalised for eye and brain lesions during the war with Vietnam, could be considered indirect proof that the Russians had already reached the stage of testing radiation arms in a real conflict while the Americans were still talking about them. Alarmed by such information concerning Soviet progress in the field of arms based on 'non-conventional' technology, the Americans overcame their initial scepticism and stepped up research, both in the field of high energy physics and in the field of charged-particles, so as not to lose this particular race in which so much is at stake: the ultimate objectives go way beyond battles between satellites and could even be said to be part of a design for world hegemony.

Both CPB and high-energy laser weapons could theoretically be used to intercept ground-launched ICBMs, or SLBMs (Submarine Launched Ballistic Missiles), with nuclear warheads, and destroy them in space.

A further development in the technology of particle physics involves an even more powerful device which could be used to generate particle beams from spacecraft and transmit extremely high levels of radiation to Earth with effects similar to those produced by neutron bombs. In other words, with suitable power levels, charged particles could be propagated through the atmosphere to produce a radiation cone which would have lethal effects on populated areas.

On the basis of experience acquired in the meanwhile, the Americans have decided to carry out a crash programme in an attempt to make up for lost time. Plans drawn up by each of the armed forces have been examined on the basis of which a unified plan has been laid down, which has two main directions: one, conventional weapons, and the other, newly-conceived systems.

Regarding the former, satellite-killers are to be developed with similar capabilities to those tested by the Russians. One such system involves a series of small self-propelled mini-satellites, ejected from another spacecraft and guided onto their targets by an on-board IR seeker which would exploit the difference in temperature between the metal of the satellite and the surrounding space. Another development for self-defence from space attacks would be to increase the manoeuvrability of satellites or ICBMs to enable them to avoid interception (MaRV: Maneuverable Re-entry Vehicles).

Weapons of new conception are those which utilise new, very advanced technology related to high-energy physics and particle physics. High-energy laser weapons and CPB weapons are grouped together under the heading of 'radiation arms' or 'directed-energy arms'. The ultimate objective of the US programme for the development of such weapons, called the 'Talon Gold Programme', is to create a defence against ballistic missiles in space using high-energy lasers which would be installed on spacecraft or space stations. Research in this new field is carried out at the Lawrence Livermore Laboratories in California and at Los Alamos in New Mexico.

In the summer of 1977, it was officially announced in the United States that a high-energy laser weapon had for the first time destroyed a missile-target (a NIKE-Hercules) in flight; the weapon used was a fluoride deuterium laser emitting very high power IR energy at a wavelength of 3.8 microns. Nevertheless, the Americans were well aware that the Russians were about ten years ahead of them also in the use of high-energy laser weapons in space. In an attempt to bridge this dangerous gap, they embarked on a programme of research and development of suitable electronic and electro-optical counter-measures capable of neutralising these new radiation weapons, the mere possession of which was enough to upset the balance of military power, both strategic and tactical.

Plans for the development of radiation weapons drawn up by the US armed forces followed different paths according to the specific sector of possible military applications of such weapons. Besides their potential use against ICBMs and satellites, plans have also been made for their use against landmines, torpedoes, attack and strike aircraft and tanks. The CPB must concentrate enough energy on its target to detonate the high explosive in a nuclear warhead, torpedo warhead or landmine; against metal targets, such as aircraft, spacecraft, satellites and tanks, the Americans aim to develop a CPB which would produce enough heat to destroy all the electronic equipment on board immediately, and then, as the weapon closed up to the target, also seriously damage the metallic structure itself.

While the Americans were struggling against the many bureaucratic difficulties which hindered the development of their programme, the Soviet Union reached a milestone in its work on radiation weapons. In September 1979, an electron beam was tried out against various military targets, including an ICBM, solid materials and high explosive, with complete success. These tests, which were carried out near Leningrad, may turn out to have been the prelude to operational use on the battlefield of the prototype of a weapon exploiting such a beam.

Besides the technical and industrial difficulties involved in developing these new, unconventional weapons, there is also the problem of transporting them into space, given their huge dimensions and enormous weight. With the launching and subsequent re-entry of the US Space Shuttle *Columbia* in mid-April 1981, the Americans took a giant step forwards in solving this problem in particular, as well as, more generally, in the 'space race' with the Soviet Union.

Re-usable spacecraft like the *Columbia* have a great load capacity and, besides being able to transport laboratories, telescopes and satellites of all kinds, they can also carry out several military functions. They can be used to transport heavy radiation weapons, such as high-energy lasers and CPBs, intended to destroy enemy satellites or ballistic missiles, and powerful EW equipment, such as jammers and deception jammers, capable of blinding enemy surveillance satellites or deviating enemy ICBM in the event of war. Confirmation of the potential military use of the Space Shuttle by the Americans is provided by the fact that at least twenty-one of the sixty-eight missions scheduled for these spacecraft have been classified by the Pentagon as top secret.

In March 1983, the US President, Ronald Reagan, in his famous 'Star Wars' speech, officially announced a new Defence doctrine based on space-age weaponry. He said that the United States would abandon the old strategy of détente achieved through the threat of massive nuclear retaliation and would pursue a new strategy based on the ability to prevent nuclear war. It would be a defensive strategy employing weapons designed to intercept and destroy incoming enemy missiles. These weapons would be 'directed-energy' weapons—high-energy lasers in particular. Since the technology needed for such a strategy does not yet exist, he appealed to the North American scientific community to dedicate their efforts to the creation of an anti-missile defence system which would render nuclear weapons impotent and obsolete.

According to experts, this project will involve deploying eighteen space stations into orbit, each equipped with high-energy lasers and revolving in three polar orbits; it could, they opined, be put into operation during the 1990s. If successful, this project would provide the means to neutralise a mass attack of enemy ICBMs launched from any point on Earth. The designed operational range would be 5000 miles. Each station would be able to direct approximately 1000 laser-pulses on as many targets and would comprise a sophisticated target detection and acquisition system, a very high power laser, a large mirror to focus the laser beam onto the targets and a target aiming and tracking system.

For each shot, the laser will emit a power of approximately 10 MW (Megawatts) for a period of only 10 seconds. Lasers developed so far have achieved 2 MW and projects for 5 MW are underway. In the next twenty years, when new lasers and powerful electronic beams pass from the research to the developmental stage, it seems likely that powers of the order of 10 MW will be feasible. However, the major problem is aiming and focussing the beam on target as the accuracy required is 1 metre from a range of 10,000 kms. This laser beam would require orbiting mirrors 10 metres in diameter and the development of sophisticated microwave search systems and laser aiming systems. A high technology programme is already underway in the United States with the purpose of solving this problem.

In April 1983, the Space Shuttle *Challenger*, the second of four operational 'Orbiters' was sent into orbit. It transported a TDRS-A (Tracking and Data Relay Satellite) which was launched some days later. During the mission, extravehicular activity lasting 3.5 hours was carried out by two members of the crew on the fourth day.

In February 1984, *Challenger*'s crew gave a successful demonstration of NASA's Manned Maneuvering Unit (MMU). The untethered spacewalks made by US astronauts Bruce McCandless and Bob Steward, who each used the MMU on two separate occasions, demonstrated that it is possible for men to approach enemy surveillance satellites for the purpose of destroying them or degrading their performance. The fantastic achievement of the two US astronauts has opened a new frontier in what human kind can do in space and paved the way for many important operations in future space Electronic Warfare.

NASA have stated that they plan to have a permanent space station operating by 1991. This manned base will have a crew of six to eight persons, with computers controlling each task. It would evolve over many years and see many cycles of technology and utilisation. One such utilisation will be the setting up of an Electronic Warfare Command and Control Centre.

For most of us, space stations belong to the realm of science fiction, exemplified, perhaps, by memorable sequences from the film *Star Wars*, but such fantastic things are fast becoming a reality. The Superpowers are already studying future electronic combat in space and 1991 is not far away. The era of space fiction is over; it has become a reality and a sort of electronic 'star wars' could be what the future holds in store—a Space Shuttle fleet, fitted with high-energy laser weapon systems patrolling the 'skies' ready to intercept and destroy enemy ICBMs still in their booster phase. The Russians, as has been discussed, have also been devoting great efforts to developing technology related to 'directed-energy' weapons and a recent US secret intelligence report on the Soviet laser programme stated that the Soviet Union would be able to deploy a space-based high-energy laser weapon station as early as 1988.

The biggest drawback to orbiting laser weapons seems to be their vulnerability to countermeasures. It is fairly easy to envisage how to jam a space acquisition or aiming system even with today's technology. A couple of stations put out of action either by failure of their equipment or enemy action would nullify the effectiveness of the whole system and remove the 'space-umbrella', allowing enemy ICBMs to rain down.

There are two kinds of countermeasures applicable in space warfare: countermeasures against the platforms or space-stations (Shuttle, Soyuz, satellites, etc.) and countermeasures against 'directed-energy' weapons. Both require threat warning receivers for immediate detection of enemy radar, laser or IR source (booster, exhaust, etc.). Against the platforms, similar ECM equipment to that used on Earth could be employed: onboard jammers and expendable jammers, chaff, IR flares, radar absorbing shields, and so on. Against the laser beam, laser decoys, mirrors and space mines could be used—or any other electro-optical countermeasures (EOCM) which may emerge from technological progress. Thus, a sophisticated laser antimissile system of astronomical cost could be put out of action by countermeasures costing much less. However, it is likely that, to compensate for this inherent weakness in electro-optical (EO) weapons, their vulnerability to EOCM, efforts will be devoted to finding effective EO counter-countermeasures (EOCCM).

It is not now so farfetched to suppose that one Superpower, were it to acquire the ability to destroy enemy satellites and ICBMs in space before the other and thus become virtually invulnerable to a pre-emptive nuclear strike, might be induced to launch a nuclear attack against the other and destroy him.

Apart from this pessimistic hypothesis, it is, nevertheless, likely that, in future international crises, space will provide the perfect arena for a show of strength by the most technologically advanced Superpowers in these new fields of military art and the connected branches of applied science. In other words, a challenge could be initiated in space between spacecraft, satellites, ICBMs and 'radiation-weapons' in which the Superpower in possession of the more effective 'radiation-weapons' could destroy all enemy weapons and equipment, thus proving their potential to destroy the enemy on Earth also. In this way, without killing people or violating territorial borders, a crisis could be resolved in favour of the Superpower in possession of what many people today consider to be the 'absolute weapon', capable of winning any conflict.

However, it is unlikely that either the high-energy laser or the CPB weapon will prove to be the 'absolute weapon'. The idea of an 'absolute weapon' has always been a myth and is likely to remain so since, as the events described in this book have shown, as soon as a new weapon is developed, appropriate countermeasures are immediately devised to engage it and neutralise its effectiveness.

The classic struggle between the lance and shield, the gun and armour, the missile and electronic countermeasures will no doubt continue between radiation weapons and radiation countermeasures and between these countermeasures and relative counter-countermeasures and so on *ad infinitum*. Such is the nature of Electronic Warfare.

Endnotes

Chapter 1
1 Radio transmitter using as its radio-frequency source the oscillatory discharge of a capacitor across a coil and a spinterometer.

Chapter 4
1 Ships equipped with search radar positioned ahead of the main formation to maximise the formation's radar coverage.

Chapter 5
1 Sir Winston S. Churchill, *Their Finest Hour*, pp 381–2. Houghton Mifflin Company. Boston, UK, 1949.

Chapter 12
1 For further information on 'Ultra' and 'Enigma' see *Ultrasecret* by Winterbotham and *Il vero traditore* by A. Santoni, published by Mursia.

Chapter 13
1 Radar capable of continuously measuring the distance, bearing and elevation of a target in order to determine its future point to enable missiles or artillery to home in on the target at that point.
2 Tactical Air Navigation equipment, a military radio air navigation system in which a UHF ground transmitter emits signals which interrogate equipment installed on the aircraft, which responds with signal data regarding the aircraft's direction and distance from the station.

Chapter 14
1 In ordinary language, the word 'intelligence' means 'raw information'; in technical jargon, it means the 'end product' of the processing and evaluation of data.

Chapter 15
1 The tabulation of deployment of all radars and electronic warfare systems with all their relative operative and technical characteristics.

Chapter 19
1 Tesla gave his name to the unit of magnetic induction in the International System of Measures (SI): 1 Tesla = 1 Weber/m^2.
2 The electrical activity of the human brain is normally measured, for diagnostic purposes, by an electroencephalogram and consists of sinusoidal oscillations at a medium frequency of 10 Hz with an amplitude of 10–50 microvolts. Electrical brain stimulation techniques have been studied by E. Hitzig and G. Fritsch.
3 See *Electromagnetic Field Effects*, ed. Persinger, published by Plenum Press, New York and London, 1974.

Chapter 21

1 Type 42 destroyers were equipped with terminals of the 'Scot Skynet' system for communications via satellite.

2 It has been suggested that the ship's ESM equipment was also switched off to facilitate radio communications via satellite (see *Defense Electronics*, November 1983: 'Falklands'). However, it is inconceivable that a ship acting as a radar picket would have had both its radar and ESM equipment switched off simultaneously.

3 These new generation weapons, such as Seaguard, Goalkeeper, Phalanx, etc., have very high rates of fire – up to 4000 rounds per minute. Their ammunition is also of an extremely advanced 'kinetic' type, guaranteeing destruction of a missile by one round.

4 A few weeks previously, Argentinian naval aviation had begun intensive operational testing for the employment of the Super Etendard-Exocet combination. The British-designed and built Type 42 destroyers *Hercules* and *Santissima Trinidad* were used for this purpose. It was concluded that it was possible to get close enough to Typ 42s, to which class *Sheffield* belonged, to launch Exocet missiles by low flying to avoid detection by their radar. The missiles could be launched from a distance outside the range of Sea Dart missiles and, once the missiles had been launched, the Super Etendards had nothing more to do than return to base.

5 The 'warners' of most NATO ships are, generally, programmed only for Soviet radars. Moreover, the Abbey Hill equipment had an extremely limited capacity for pre-programmed warnings. It should also be noted that the surface-to-surface (MM-38) version of the Exocet is installed on many ships of the Royal Navy.

6 Agave is a light, multi-purpose monopulse radar installed on French Super Etendard and Jaguar aircraft. Its main modes are air-to-surface and air-to-air search and tracking, data feed to the missile's radar guidance system, ranging (distance measuring) and navigation. It operates in the I-band (8–10 GHz).

7 Built by Texas Instruments during the Vietnam war, this family of smart, or laser-guided bombs (LGBs) includes various types, all equipped with the same type of guidance unit, fitted to the nose.

8 The British also declared that they had lost only twelve aircraft, of which nine were Harriers and Sea Harriers.

9 The ALQ-101 is the head of a family of pod-contained jammers. It was installed on attack aircraft used in Vietnam. Various technical improvements were made, culminating in the production of a new version, the ALQ-119, which has been sold to many countries. In the Falklands, it was installed on Vulcan bombers to neutralise the radars used to guide Argentinian Roland missiles.

Chapter 23

1 As usually happens with heavenly bodies, such as meteorites, an artificial satellite re-entering the Earth's atmosphere at very high speed is consumed by heat caused by friction with the air and burns up in the high strata of the atmosphere. This occurs at altitudes of not less than 80–90 kms. However, it can happen that a satellite, because of its considerable mass and the solidity of some of the materials of which it is composed, does not disintegrate completely during its fall through the atmosphere, with the result that fragments might fall to Earth. This was the case with the Soviet Cosmos 954, parts of which fell in Canada, and the US Skylab which fell into Australian waters.

2 An electron-volt is the amount of energy acquired by an electron when it passes from the negative to the positive pole of a 1-volt battery. One GeV equals one billion electron-volts.

Abbreviations

AA;AAA	Anti-Aircraft Artillery
AAM	Air-to-Air Missile
ADF	Automatic Direction-Finder
ADP	Automatic Data Processing
AEW	Airborne Early Warning
AGI	Auxiliary, Intelligence Gatherer (USSR)
AGM	Air-to-Ground Missile
AIM	Air Intercept Missile
AJ	Anti-Jamming
ALASCA	All-Aspect Capability (missile)
AN/ALQ, /ALR,/TPS	Joint Electronics Designation System (JETDS) numbered equipment (AN)
ARM	Anti-Radar (radiation) Missile
ARP	Antenna Rotation Period
ASAT	Anti-Submarine Warfare
ASDIC	Anti-Submarine Detection Investigation Committee (Sonar)
ASV	Anti-Surface Vessel radar
ASW	Anti-Submarine warfare
AWACS	Airborne Warning and Control System
CEP	Circular Error Probability
C³I	Command, Control and Communications Intelligence
C³CM	C³ Counter Measures
CIA	Central Intelligence Agency
CIC	Command Information and Control centre
cm	Centimetre
COMINT	Communications Intelligence
CPB	Charged Particle Beam
CW	Continuous Wave (radar)
DF	Direction-Finder
DJ	Deception Jamm(er/ing)
ECCM	Electronic Counter-Countermeasures
ECM	Electronic Countermeasures
EHF	Extremely High Frequency
ELF	Extremely Low Frequency

ELINT	Electronic Countermeasures
EOCCM	Electro-Optic Counter-Countermeasures
EOCM	Electro-Optic Countermeasures
ESM	Electronic Support Measures
EW	Electronic Warfare; Early Warning
FLIR	Forward Looking Infrared
FM	Frequency Modulation
GHz	1 GHz = 1000 MHz
GRU	*Gosurdarstarvenoi Razvedyvatelnaya* (State Military Information Agency, USSR)
HF	High Frequency
HMS	His/Her Majesty's Ship/Submarine
HOJ	Home-on-Jam
HUD	Head-Up Display
Hz	Hertz
ICBM	Inter-Continental Ballistic Missile
IFF	Indentification Friend or Foe
ILS	Instrument Landing System
INS	Inertial Navigation System
IR	Infrared or Infrared Radiation
IRBM	Intermediate Range Ballistic Missile
IRCM	Infrared Countermeasures
IRSM	Infrared Support Measures
IRWR	Infrared Warning Receiver
Kc/s	Kilocycles per second
KGB	*Komitet Gosurdarstarvenoi Bezopasnost* (Committte for State Security, USSR)
km; km/h	kilometre; kilometre per hour
LADAR	Laser Detection and Ranging
LLTV	Low-Light-Level Television
LORAN	Long-range navigation
LORO	Lobe-on-Receive Only
m.; mm	metre; millimetre
Mc/s	Megacycle per second

MHz	Megahertz
mph	miles per hour
MRBM	Medium Range Ballistic Missile
MW	Megawatts (10^6W)
NATO	North Atlantic Treaty Organisation
NDB	Non-Directional Radio Beacon
nm	nanometre (10^{-9}m)
NSA	National Security Agency
PAM	Pulse Amplitude Modulation
PCM	Pulse Code Modulation
PD	Pulse Doppler
PDM	Pulse Duration Modulation
PFM	Pulse Frequency Modulation
PRF	Pulse Repetition Frequency
RAF	Royal Air Force
RAM	Radar Absorbent Material
RDT	Radio-Detector Telemeter
RPV	Remotely-Piloted Vehicle
RWR	Radar Warning Receiver
SALT	Strategic Arms Limitation Treaty
SAM	Surface-to-Air Missile
SIGINT	Signal Intelligence
SLAR	Side-Looking Airborne Radar
SLBM	Submarine Launched Ballistic Missile
SONAR	Sound Navigation and Range
TACAN	Tactical Air Navigation
TOW	Tube-launched, Optical-tracking, Wire-guided missile
TWS	Track-While-Scan
UHF	Ultra High Frequency
US	United States
USA	United States of America
USAF	United States Air Force
VHF	Very High Frequency
VLF	Very Low Frequency

Glossary

ACTIVE ELECTRONIC COUNTERMEASURES The impairment of enemy electronic detection, control or communications devices/systems through deliberate jamming or deception.

ACTIVE JAMMING Intentional radiation or reradiation of electromagnetic waves with the object of impairing the use of a specific portion of the electromagnetic wave spectrum.

ANTI-JAMMING (AJ) The technique of minimising the effect of enemy electronic countermeasures to permit echoes from targets detected by radar to be visible on the indicator. A synonym for electronic counter-countermeasures. *See* ECCM.

CHAFF An airborne cloud of lightweight reflecting objects typically consisting of strips of aluminium foil or metallic coated fibres which produce clutter echoes in a region of space.

COMINT Communications Intelligence.

COMMUNICATIONS COUNTERMEASURES Electronic countermeasures used specifically against communications.

COMMUNICATIONS DECEPTION Use of devices, operations and techniques with the intent of confusing or misleading the use of a communications link or a navigation system.

COMMUNICATIONS JAMMING The part of electronic jamming used against a medium that employs electromagnetic radiation to convey information from one person or headquarters to another.

CONFUSION REFLECTOR Reflector of electromagnetic radiation used to create echoes for confusion against radar, guided missiles and proximity fuses. *See* Chaff.

DECOY A device or devices used to divert or mislead enemy defensive systems so as to increase the probability of penetration and weapon delivery

ELECTRONIC COUNTER-COUNTERMEASURES (ECCM) That division of electronic warfare involving actions taken to ensure friendly effective use of the electromagnetic spectrum despite the enemy's use of electronic warfare.

ELECTRONIC COUNTERMEASURES (ECM) That division of electronic warfare involving actions taken to prevent or reduce an enemy's effective use of the electromagnetic spectrum.

ELECTRONIC DECEPTION The deliberate radiation, reradiation, alteration, absorption or reflection of electromagnetic radiations in a manner intended to mislead an enemy in the interpretation of or use of information received by his electronic systems. There are two categories of electronic deception—
1. **Manipulative deception**: The alteration or simulation of friendly electromagnetic radiations to accomplish deception. 2. **Imitative deception**: The introduction of radiations into enemy channels which imitate his own emissions.

ELECTRONIC INTELLIGENCE (ELINT) The intelligence information product of activities engaged in the collection and processing, for subsequent intelligence purposes, of foreign, non-communications, electromagnetic radiations emanating from other than nuclear detonations or radioactive sources.

ELECTRONIC JAMMERS 1. **Expendable**: A transmitter designed for special use such as dropped behind enemy lines. 2. **Repeater**: A receiver-transmitter device which, when triggered by enemy radar impulses, returns synchronised false signals to the enemy equipment. The returned impulses are spaced and timed to produce false echoes or bearing errors in the enemy equipment.

ELECTRONIC JAMMING The deliberate radiation, reradiation, or reflection of electromagnetic energy with the object of impairing the use of electronic devices, equipment or systems being used by the enemy.

ELECTRONIC ORDER OF BATTLE A listing of all the electronic radiating equipment of a military force giving location, type, function and other pertinent data.

ELECTRONIC RECONNAISSANCE The detection, identification, evaluation and location of foreign, electromagnetic radiations emanating from other than nuclear detonations or radioactive sources.

ELECTRONIC WARFARE Military action involving the use of electromagnetic energy to determine, exploit, reduce or prevent hostile use of the electromagnetic spectrum and action which retains friendly use of the electromagnetic spectrum. There are three divisions within electronic warfare: electronic warfare support measures (ESM), electronic countermeasures (ECM) and electronic counter-countermeasures (ECCM).

ELECTRONIC WARFARE INTELLIGENCE (ELINT) Electronic Warfare intelligence is the product resulting from the collection, evaluation, analysis, integration and interpretation of all available information concerning foreign nations or areas of operations which are significant to electronic warfare.

ELECTRONIC WARFARE SUPPORT MEASURES (ESM) That division of electronic warfare involving actions taken to search for, intercept, locate and identify immediately radiated electromagnetic energy for the purpose of immediate threat recognition. Thus, electronic warfare support measures provide a source of information required for immediate action involving electronic countermeasures, electronic counter-countermeasures, avoidance, targetting and other tactical employment of forces.

ELECTRO-OPTIC COUNTER-COUNTERMEASURES (EOCCM) Actions taken to ensure the effective friendly use of the electro-optic spectrum despite the enemy's use of countermeasures in that spectrum.

FERRET An aircraft, ship or vehicle especially equipped for the detection, location, recording and analysing of electromagnetic radiations.

FLARE In the countermeasure sense, a flare is a pyrotechnic target launched from an aircraft or other vehicle causing infrared homing missiles or other optical devices to be decoyed away from the true target.

FREQUENCY AGILITY This term refers to a radar's ability to change frequency within its operating band.

HOME-ON-JAM (HOJ) A method of passive guidance designed to use the jamming signal emitted by the target to track the target in angle.

313

INFRARED COUNTER-COUNTERMEASURES Actions taken to effectively employ friendly infrared radiation equipments and systems in spite of the enemy's actions to counter their use.

INFRARED COUNTERMEASURES 1. Countermeasures used specifically against enemy threats operating in the infrared spectrum. 2. Actions taken to prevent or reduce the effectiveness of enemy equipment and tactics employing infrared radiation.

INTERCEPT RECEIVER Special calibrated receiver which can be tuned over a wide frequency range to detect and measure radio signals transmitted by the enemy. Also called Search Receiver.

NAVIGATION COUNTERMEASURES The detection and evaluation of enemy electronic aids to navigation, and the use of jamming and deception to interfere with enemy use of such aids.

OPTICAL COUNTERMEASURES (OCM) 1. Applications of electronic countermeasures in the visible light portion of the electromagnetic spectrum. 2. Actions taken to prevent or reduce an enemy's effective use of the visible spectrum. Also called Visual Countermeasures.

PASSIVE ELECTRONIC COUNTERMEASURES Electronic countermeasures based on the reflection, absorption or modification of the enemy's electromagnetic energy. This is not currently used but it is based on the presence or absence of an electronic transmitter.

RADAR ABSORBENT MATERIAL (RAM) Radar absorbent material used as a radar camouflage device to reduce the echo area of an object i.e. an aircraft.

RADAR DECOY Reflecting object used in radar deception, having the same reflective characteristics as a target.

RADAR SILENCE An imposed discipline prohibiting the transmission by radar of electromagnetic signals on some or all frequencies.

RADAR WARNING RECEIVER (RWR) A wideband crystal video receiver (usually airborne) designed to intercept, identify and display the type of, and direction to radar emitters. It is not a homing system.

RADIO DECEPTION The employment of radio to deceive the enemy. Radio deception includes sending false enemy call signs.

ROPE An element of chaff consisting of a long roll of metallic foil or wire which is designed for broad, low frequency response.

SIGNAL INTELLIGENCE (SIGINT) A generic term which includes both communications intelligence and electronic intelligence.

WINDOW 1. Strips of frequency-cut metal foil, wire or bars which may be dropped from aircraft or missiles, or expelled from shells or rockets as a radar countermeasure. A confusion reflector. 2. A passive radar deception or confusion device, usually consisting of some metallic structure, designed in size and shape to effectively reflect impinging signals, so as to simulate a true target.

Select Bibliography

ARENA, N.: *Il radar.*

BARKLEY, G.: *La guerre des services secrets.*
BEKKER, C: *Luftwaffe – Radar-Duell in Dunkel.*
BENNET, G.: *Naval Battles of the First World War.*
BERGIER, J.: *L'espionage scientific.*
BOSCHESI, B. P.: *Le grandi battaglie segrete della II Guerra Mondiale.*
BRANDT, E.: *La capture de L'USS* Pueblo.
  The last voyage of USS Pueblo.
Brassey's Artillery of the World, Brasseys, 1977.
BROWN, A.C.: *Bodyguard of Lies.*
  Carrier Operations in World War Two.
BUSHBY, J. R.: *Air Defence of Great Britain.*
BURNS, M. G.: *McDonnell Douglas F-4K and F-4M Phantom II*, Osprey Publishing, London, 1984.

CARROLL, J. M.: *Secrets of Electronic Espionage.*
CHURCHILL, SIR WINSTON S.: *World War II.*
COOK. C. AND STEVENSON, J.: *The Atlas of Modern Warfare.*
COLLIER, B.: *The History of Air War.*
COX, J.: Overkill: *The Story of Modern Weapons*, Kestrel Books, UK, 1976.

DE ARCANGELIS, ADM. M.: 'Air Superiority and electronic superiority', in *Aviation & Marine*, December 1976, pp 63–6. 'Electronic warfare', in *Aviation & Marine*, No. 48, 1977, pp 63–6; No. 49, 1978, pp 63–6; No. 50, 1978, pp 63–6 'Electronic warfare and land battles', in *Armies & Weapons*, No. 39, 1977, pp 59–62; No. 40, 1978, pp 63–6; No. 41, 1978, pp 53–6.
DE LAGE: *Le drame du Jutland.*

EUSTACE, H. F.: *The International Countermeasures Handbook*, editions 1975–79.
  'Airborne equipment: a mature military asset', *International Defence Review*, February 1976.
'Army EW: Current US Concepts', in *International Defence Review*, June 1976.
'A US View of Naval EW', in *International Defence Review*, April 1976.
'Js E.W. Still a Pentagon Stepchild', in *Electronic Warfare*, July/August 1976.
Frank's Report, HMSO, London, 31 December 1982.

GEISENHEYNER, S. J.: 'Electronic Warfare: a Battle of Wits', in *Aerospace International*, January/February 1973.
GENTLE, E. J. AND REITHMAIER, W.:*Aviation/Space Dictionary*, Aero Publishing Inc., Fallbrook, California, USA, 1980 (6th edn).
GORDON, D.E., *Electronic Warfare*, Pergamon Press, Oxford, 1981.
GUNSTON, W. T. (BILL): *F-4 Phantom*, Ian Allan Ltd, Shepperton, Surrey, 1977.
  General Dynamics F-111, Ian Allan Ltd, Shepperton, Surrey.

HACKETT, GEN. J. AND OTHERS: *The Third World War.*
HARRIS, D. B., LORENSEN, H. O., AND STIBER, S.: *History of Electronic Countermeasures.*

HARTMAN, R.: 'The Real EW Pioneer' (Dr F. E. Terman), *Military Electronics Countermeasures*, February 1978.
HECHT, J., 'Laser Weapons: A Status Report', in *Analog 97*, No. 10, 1977.
Her Majesty's Government White Paper, Command 8758, *The Falklands Campaign: The Lessons*, HMSO, London, 1982.

HERZOG, GEN. C.: *The War of Atonement*.
HEZLET, A.: *Electronics and Sea Power*.
HOERZLING: *The Great War at Sea*.
HOUGH, R.: *La flotta suicida*.
HURST, M.: *Airborne Early Warning*, Osprey Publishing, London, 1983.

JACKSON, R.: *Air War Over Korea*.
JAFFRELOT: 'La guerre des sorciers 1939–1945', in *La Revue Maritime*.
Jane's All the World's Aircraft, Jane's Publishing London, biennial.
Jane's all the World's Fighting Ships, Jane's Publishing, London, biennial.
Jane's Communications Systems, Jane's Publishing, London, biennial.
Jane's Weapons Systems, Jane's Publishing, London, biennial.
JONES, R. V.: *Most Secret War*.

KAMSTRA, J.: 'History of ECM' in *Countermeasures*, June/July 1975, 1976, 1977.

LANGFORD, D.: *War in 2080: The Future of Military Technology*, Westbridge Books (David & Charles), Newton Abbot, Devon, 1979.
LATOUR, C.: 'La guerra elettronica', in *Nato's Fifteen Nations*, April/May 1974.

MARRIOTT, J.: 'New weapons for defence in Europe', in *Nato's Fifteen Nations*, December 1973, January 1974.
MCKEE, A.: *Bataille de la Manche*.
MENGERS, P.: *Applications and principles of Infrared Technology*.
MILLER, B.: 'US exploits spurious radiation' (of internal combustion engines), in *Aviation Week & Space Technology*, January 1972.

NOVIKOV-PRIBOI, A.: *La tragédie de Tsushima*.
NUNZ, G.: *Electronics in Our World*.

O'BALLANCE, E.: *The Electronic War in the Middle East, 1968–70*.
The Wars in Vietnam.

PEILLARD, L.: *The Battle of the Atlantic*.
PEEBLES: *War in Space*, Blandford Press, Poole, Dorset, UK, 1983.
PERSINGER, editor: *Electromagnetic Field Effects*, Plenum Press, New York and London, 1974.
PESCI, G.: *La Guerra attraverso l'etere*.
POTTER, J. D.: *Fiasco – The Break-out of the German Battleships*.
PRICE, A.: *Aircraft versus Submarine*, Naval Institute Press, Annapolis, USA.
Battle over the Reich, Ian Allan Ltd, Shepperton, Surrey.
Electronic Warfare, MacDonald and Jane's Ltd, London, 1977.
Instruments of Darkness, MacDonald and Jane's Ltd, London.
The Bomber in World War II, MacDonald and Jane's Ltd, London, 1976.

RAMPANTS: *US Electronic Espionage: A Memoir.*
ROSS, T. AND WISE, D.: *The U-2 Affair.*

SANTONI, A.: *Il vero traditore,* Mursìa, Milan, Italy.
SCHLESINGER, R. J. AND OTHERS: *Principles of Electronic Warfare,* 1961.
SCHOFIELD, B. B.: *La fine della* Bismarck.
SHEPHERD, G.: *German Aircraft of World War II.*
SHORES, C. F.: *Mediterranean Air War.*
SIMS, E. H.: *Sfide nei cieli.*
SUNDARAM G. S.: 'Electronic Warfare at Sea', in *International Defence Review,* April
   1976. 'Expendables in Electronic Warfare: Proven Decoys for survival' in
   International Defence Review, December 1976. 'EW Dedicated Aircraft: The EA-
   6B and EF-111A Systems., in *International Defence Review,* February 1977.

TIBERIO, V.: *Introduzione alla tecnica radio e radar.*

ULSAMER, E.E.: 'Electronic Warfare and Smart Weapons', in *Aerospace International,*
   July/August 1976.

University of Tel Aviv Publishing Project: *Military Aspects of the Israeli-Arab
   Conflict.*

WINDROW, M. C. editor: *Aircraft Profiles Nos 133–168,* Profile Publications Ltd,
   Windsor, UK.
WINTERBOTHAM: *Ultrasecret.*
WISE, D.: *The U-2 Affair.*

Periodicals and Journals

The following are among those which carry articles and series relating
   to aspects of Electronic Warfare:

Analog
Aerospace International
Armies and Weapons
Aviation and Marine
Air International
Aviation Week & Space Technology
Born in Battle
Defence Electronics
International Defence Review
Interavia
La Revue Maritime

Index